Advances in Fluid Mechanics and Turbomachinery

Springer

Berlin
Heidelberg
New York
Barcelona
Budapest
Hong Kong
London
Milan
Paris
Santa Clara
Singapore
Tokyo

Hans J. Rath · Christoph Egbers (Eds.)

Advances in Fluid Mechanics and Turbomachinery

With 138 Figures

Springer

Professor Dr. Hans J. Rath
Dr.-Ing. Christoph Egbers
Universität Bremen
Zentrum für angewandte Raumfahrt-technologie und Mikrogravitation
Am Fallturm
D-28359 Bremen

ISBN-13:978-3-642-72159-5

Library of Congress Cataloging-in-Publication Data

Die Deutsche Bibliothek – CIP-Einheitsaufnahme
Advances in fluid mechanics and turbomachinery / Hans J. Rath ; C. Egbers (ed.). - Berlin ; Heidelberg ; New York ; Barcelona ; Budapest ; Hong Kong ; London ; Milan ; Paris ; Santa Clara ; Singapore ; Tokyo : Springer, 1998
ISBN-13:978-3-642-72159-5 e-ISBN-13:978-3-642-72157-1
DOI: 10.1007/978-3-642-72157-1

Softcover reprint of the hardcover 1st edition 1998

Cover Design: de'blik, Berlin
Typesetting: Camera-ready by Editors
SPIN: 10669076 62/3020-5 4 3 2 1 0 - Printed on acid -free paper

Preface

This special volume contains several articles written in honour of Professor Ingolf Teipel on the occasion of his 65th birthday.

Professor Teipel was born on March 15th, 1933 in Wissen/Sieg. He started his school career in Betzdorf/Sieg. From 1943 to 1951 he went to the Gymnasium in Betzdorf, where he obtained his school-leaving exam after only 8 years. In 1951 Professor Teipel started his studies in Mechanical Engineering at the Technical University of Aachen and got his diploma at the Aerodynamic Institute in 1956. Then Professor Teipel decided to join the research group of the Institute of Theoretical Gasdynamics at the Deutsche Versuchsanstalt für Luft- und Raumfahrt (DVL), the famous institute of Professor Oswatitsch, where he spent about 13 years until 1968.

At the Technical University of Aachen he finished his dissertation devoted to spherical-symmetric shock waves and he obtained his doctorate in Mechanical Engineering in 1960. Some years later he went as a guest scientist to the Aerospace Laboratories of Dayton, Ohio, USA. In 1965 Professor Teipel finished his habilitation work with a paper on unsteady transonic pressure forces at the Technical University of Aachen, where he recieved the venia legendi in fluid mechanics. In this year, Professor Teipel was given the freedom of the state Tennessee, USA.

In 1968 he changed his place of work and he followed an offer as a Professor at the Institute of Mechanics at the Technical University of Hannover, where he was responsible for the Fluid Dynamics Department. In 1969 Professor Teipel followed an offer as guest professor at the University of Strathclyde in Glasgow, Scotland. In 1987 Professor Teipel went to Tokyo Denki University, Japan, as a visiting professor.

Many students and co-workers gained from Professor Teipels excellent ability as a teacher and as a scientist. Most of them owe their successful career in industry and science to him.

Some of his friends have expressed their best wishes and respects which are contained in this volume in the form of scientific contributions. The papers in this volume are mostly in the area of computational fluid dynamics (CFD), which is Professor Teipel´s favorite field of research. Furthermore, to some extend this volume contains also contributions from the field of new experimental methods and diagnostics applied to fluid dynamics, combustion and turbomachinery. The contributed papers cover diverse topics such as, pipe flows, shock tube flows, compressor flows as well as velocity and turbulence measurements of flow conditioners. There is also a survey article on recent flow computations on high performance computers. Articles are also devoted to liquid-liquid systems, rotating fluid flows and combustion diagnostics.

The editors would like to express their gratitude to all the contributors for this volume. We are also thankful to Dr. Merkle and Mrs. Grünewald-Heller and their colleagues at Springer-Verlag, Heidelberg, for their friendly cooperation in producing this volume.

Bremen, December 1997

Hans J. Rath
Christoph Egbers

Table of Contents

List of Contributors

C. Egbers
ZARM, Universität Bremen
Am Fallturm
D-28359 Bremen, Germany

Ch. Eigenbrod
ZARM, Universität Bremen
Am Fallturm
D-28359 Bremen, Germany

M. Endo
Department of Mechanical Engineering
Tokyo Metropolitan College of Technology
1-10-40 Higashi-Ohi
Shinagawa-ku
Tokyo 140, Japan

K. Gersten
Ruhr-Universität Bochum
Institut für Thermo- u. Fluiddynamik
Universitätsstraße 150 IB
D-44801 Bochum, Germany

Ch. Hartmann
TH Darmstadt
Institut für Mechanik
Hochschulstraße 1
D-64287 Darmstadt, Germany

J. Iwamoto
Department of
Mechanical Engineering
Tokyo Denki University
2-2 Kanda-Nishiki-Cho
Chiyoda-Ku
Tokyo 101, Japan

K. Kalkühler
Lehrstuhl für Strömungslehre
Universität GH Essen, FB 12
Schützenbahn 70
D-45117 Essen, Germany

I. Kassens
Universität Hannover
Institut für Strömungsmaschinen
Appelstr. 9
D-30167 Hannover, Germany

S. Kasten
TH Darmstadt
Institut für Mechanik
Hochschulstraße 1
D-64287 Darmstadt, Germany

M. Klika
Ruhr-Universität Bochum
Institut für Thermo- u. Fluiddynamik
Universitätsstraße 150 IB
D-44801 Bochum, Germany

M. Heinrich
Universität Hannover
Institut für Mechanik
Appelstr. 11
D-30167 Hannover, Germany

J. König
ZARM, Universität Bremen
Am Fallturm
D-28359 Bremen, Germany

H. G. Hornung
Graduate Aeronautical
Laboratories
California Institute of
Technology
Pasadena CA 91125, USA

E. Krause
RWTH Aachen, Institut
für Aerodynamik
Templergraben 55
D-52062 Aachen

M. Meinke
RWTH Aachen, Institut für
Aerodynamik
Templergraben 55
D-52062 Aachen

M. Rautenberg
Universität Hannover
Institut für Strömungsmaschinen
Appelstr. 9
D-30167 Hannover, Germany

W. Merzkirch
Lehrstuhl für Strömungslehre
Universität GH Essen, FB 12
Schützenbahn 70
D-45117 Essen, Germany

K.G. Roesner
TH Darmstadt
Institut für Mechanik
Hochschulstraße 1
D-64287 Darmstadt, Germany

V. Noshadi
Inst. f. Strömungslehre u.
Wärmeübertragung
TU Wien
Wiedner Hauptstraße 7/322
A-1040 Wien

O. Romberg
Universität Hannover
Institut für Mechanik
Appelstr. 11
D-30167 Hannover, Germany

K. Popp
Universität Hannover
Institut für Mechanik
Appelstr. 11
D-30167 Hannover, Germany

W. Schneider
Institut für Strömungslehre
und Wärmeübertragung
TU Wien
Wiedner Hauptstraße 7/322
A-1040 Wien, Austria

J.J. Quirk
Graduate Aeronautical Laboratories
California Institute of Technology
Pasadena CA 91125, USA

D. Raake
Lehrstuhl für Mechanik
Universität GH Essen, FB 12
Schützenbahn 70
D-45127 Essen, Germany

H. J. Rath
ZARM, Universität Bremen
Am Fallturm
D-28359 Bremen, Germany

J. Siekmann
Lehrstuhl für Mechanik
Universität GH Essen, FB 12
Schützenbahn 70
D-45127 Essen, Germany

I. Teipel
Universität Hannover
Institut für Mechanik
Appelstr. 11
D-30167 Hannover, Germany

A. Wiedermann
Takasago Research
& Development Center
Mitsubishi Heavy Industries, LTD
2-1- Shinhama Arai-Cho
Takasago, Hyogo Pref. 676 Japan

Examples of Recent Flow Computations

Egon Krause and Matthias Meinke

Aerodynamisches Institut der RWTH Aachen, 52062 Aachen, Germany
E-Mail: ek@aia.RWTH-Aachen.de

Abstract Four examples of external and internal flow simulations are presented. The first simulation reported exemplifies the computation of steady supersonic flow with a freestream Mach number of Ma=3.8 around a model of the space transportation system ELAC 1, including the chemically reacting flow in the exhaust nozzle. The flow around the configuration is predicted with a numerical solution of the parabolized Navier-Stokes equations for laminar conditions, while for the prediction of the nozzle flow the Navier-Stokes equations are solved in the symmetry plane of ELAC 1 under the assumption of chemical equilibrium. The second example demonstrates results obtained with large-eddy simulations of internal and free turbulent flows. Examples are presented for turbulent flows in bended pipes, pipe junctions and mixing processes in jets and jet apparatus. The difficulty of prescribing suitable inflow conditions for the jets was circumnavigated by obtaining velocity and density distributions from fully developed pipe flows. The third example is concerned with the simulation of the unsteady, three-dimensional, compressible flow in a cylinder of a four-valve piston engine during the suction and the compression stroke. The vortex structures generated during the intake stroke were simulated with a numerical solution on a boundary-fitted, block-structured moving grid. The fourth example shows results recently obtained in numerical simulations of a locally unsteady, incompressible swirling laminar pipe flow, bursted in a bubble-type breakdown. This flow was simulated with a dual-time stepping technique, based on the concept of artificial compressibility.

1 Introduction

Different examples of three-dimensional steady and unsteady flow problems are presented, which demonstrate the applicability of numerical methods to internal and external flow problems. The flow problems are currently investigated in various research projects. Some of the results obtained and details of the algorithms applied have already been presented recently. Hypersonic flow around ELAC with and without external combustion is discussed in [7, 12], large-eddy simulation of internal and free turbulent flows in [18, 16], flow in piston engines in [17, 15] and vortex breakdown in [11, 2]. Therefore, only the essential elements of the algorithms are given in this paper together with some new results and a summary of the major conclusions.

The examples presented require extensive computer resources in terms of storage capacity and computational work. Therefore, an efficient implementation of algorithms on high-performance parallel computers is required, which is not straightforward in general. At the Aerodynamisches Institut of the RWTH Aachen a modular program library was developed, which contains several algorithms for the solution of the Navier-Stokes equations. Currently, compressible and incompressible, steady and unsteady, two- and three-dimensional flows can be computed on block-structured curvilinear grids. The application of a standard block-structured grid system is advantageous, since it automatically provides a domain decomposition for parallelization. In addition, vector operations can easily be formulated in each domain. The various algorithms are implemented as modules in a common software environment. It is constructed in such a way that program maintenance is simplified. In addition, functions not related to a specific numerical method, like parallelized multi-block exchange or streak-line integration, are at once available for all algorithms. This program library has been ported on several vector, parallel, and vector-parallel machines.

2 Method of Solution

2.1 Governing Equations

The Navier-Stokes equations are solved for the simulation of compressible and incompressible, three dimensional and unsteady flows. In dimensionless form these equations read:

$$\bar{A} \cdot \frac{\partial \boldsymbol{Q}}{\partial t} + \frac{\partial \boldsymbol{F}_\alpha}{\partial x_\alpha} = 0 \quad . \tag{1}$$

For compressible fluid, $\bar{A}$ is the identity matrix. The vector of the conservative variables is given by:

$$\boldsymbol{Q} = J\,(\rho, \rho u_\alpha, \rho e)^T \quad .$$

ρ is the density of the fluid, u_α represents the velocity vector, and e is the internal energy. The flux vectors $\boldsymbol{E}_\alpha$ are splitted in their advective and the diffusive parts, e. g.: $\boldsymbol{E}_\alpha = \boldsymbol{E}^a - \boldsymbol{E}^d$, with

$$\boldsymbol{F}^a_\beta - \boldsymbol{F}^d_\beta = \begin{pmatrix} \varrho u_\beta \\ \varrho u_\alpha u_\beta + p\delta_{\alpha\beta} \\ u_\beta(\varrho E + p) \end{pmatrix} + \frac{1}{Re} \begin{pmatrix} 0 \\ \sigma_{\alpha\beta} \\ u_\alpha \sigma_{\alpha\beta} + q_\beta \end{pmatrix} \quad . \tag{2}$$

Herein, Re is the Reynolds-number and $\bar{\bar{\sigma}}$ is the stress tensor written as a function of the strain rate tensor $S_{\alpha\beta}$

$$\sigma_{\alpha\beta} = -2\,\mu(S_{\alpha\beta} - \tfrac{1}{3}S_{\gamma\gamma}\delta_{\alpha\beta}) \qquad \text{with} \qquad S_{\alpha\beta} = \tfrac{1}{2}(u_{\alpha,\beta} + u_{\beta,\alpha}) \quad . \tag{3}$$

For incompressible flows with constant viscosity, the Navier-Stokes equations simplify significantly. The equation for energy conservation can then be decoupled from the conservation equations for mass and momentum. The energy equation can be omitted, if the distribution of the fluid temperature is not of interest. The vector of the conservative variables in Eq. (1) is then reduced to:

$$Q = J\,(p, u_\alpha)^T \quad .$$

For incompressible flows the matrix $\bar{A}$ becomes singular, which renders the integration of the governing equations more difficult. For fluids with constant density, the vectors of the advective and diffusive fluxes reduce to, e. g.:

$$F^a_\beta - F^d_\beta = \begin{pmatrix} u_\beta \\ u_\alpha u_\beta + p\delta_{\alpha\beta} \end{pmatrix} + \frac{1}{Re}\begin{pmatrix} 0 \\ \sigma_{\alpha\beta} \end{pmatrix} \quad . \tag{4}$$

2.2 Temporal Discretization

The solution for compressible flows is integrated in time with an explicit Runge-Kutta multistep method, [10], or, alternatively with an implicit dual-time stepping method. The explicit scheme is used for unsteady compressible flow problems. The implicit dual-time stepping scheme is used for problems, for which an explicit scheme becomes inefficient, e. g. for the simulation of steady or unsteady hypersonic flows with chemical reactions.

For the simulation of incompressible flows Chorin [4] proposed to introduce an artificial equation of state, which couples the pressure- to an arbitrary density destribution in order to eliminate the singularity of the matrix $\bar{A}$. Hence, the continuity equation contains a time derivative for the pressure which vanishes for steady-state solutions and $\bar{A}$ in (1) is regular. In [3] this method was extended to unsteady flows by introducing an artificial time τ and adding a pseudo-time derivative $\tilde{A}\cdot\partial Q/\partial\tau$ to (1) such that $diag\{\tilde{A}\} = (1/\beta^2, 1, 1, 1)$, where β^2 controls the artificial compressibility. Thus, the pressure field is coupled to the velocity distribution and the governing equations can be integrated in a similar way as for compressible flows. Since a steady solution is computed within the pseudo-time τ, the additional terms vanish, and the unsteady solution of (1) at the physical time t is obtained. Details of the dual-time stepping method, the linearization, discretization and the solution of the resulting linear system of equations, are presented in [3, 20, 12].

2.3 Spatial Discretization

To preserve the conservative properties in the discretized space, Eq. (1) is formulated for a finite control volume. A corresponding difference operator, e. g. δ_ξ for the determination of the flux derivatives for a control volume in a node-centered scheme at a point (i, j, k) reads:

$$(\delta_x F)_{i,j,k} = \frac{F_{i+\frac{1}{2},j,k} - F_{i-\frac{1}{2},j,k}}{\Delta x} \quad .$$

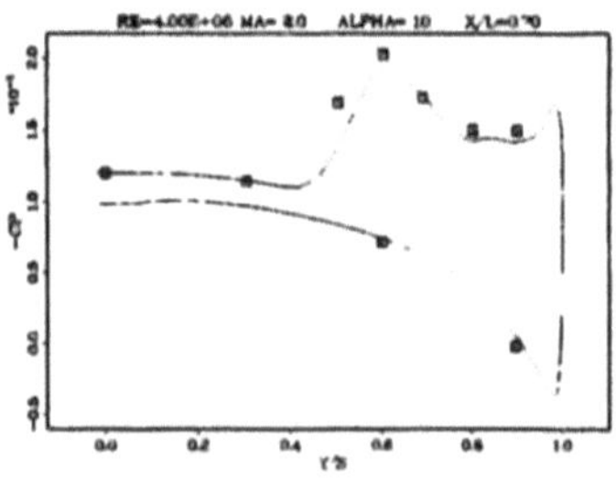

Figure2. Pressure distribution around a cross section at $x/L = 0.7$ of ELAC 1 for a Mach-number of $Ma_\infty = 4$, an angle of attack of $\alpha = 10°$ and a Reynolds-number of $Re = 4 \cdot 10^6$. Comparison of experimental data (□) and numerical solution (—).

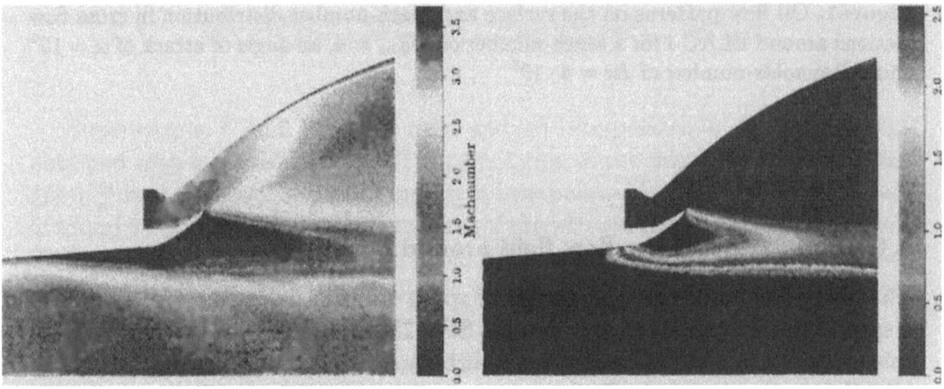

Figure3. Simulation of the flow in the symmetry plane of ELAC 1b, α=3.4°, Ma=3.8, Re=2.4 million, including chemical reactions in the SERN nozzle. Mach number distribution (left) and concentration of H_20 (right) in the exhaust plume.

angle. First simulations were carried out for the symmetry plane of ELAC 1b for a free stream Mach-number of $Ma_\infty = 3.8$, an angle of attack of $\alpha = 3.4°$ and a Reynolds-number of $Re = 2.4 \cdot 10^6$. The plume jet is simulated by assuming a stoichiometric mixture of hydrogen and air. Hydrogen is injected at the lower side of the nozzle exit to increase the thrust vector angle. The Mach number distribution and the concentration of water are presented in Fig. 3. The injection of hydrogen is clearly indicated by the high water concentration below the nozzle. An oblique shock and a flow separation is generated, which increases the thrust vector angle compared to the case without injection. These simulations are used to provide information of the overall thrust and the performance of the space transportation system. More details about this case can be found in [12].

3.2 Large-Eddy Simulation of a turbulent flows

The modelling of turbulence, especially in separated and swirling flows, is still one of the most difficult tasks in CFD. Large-eddy simulations (LES) seems to be most promising for an accurate prediction of such flow problems. In contrast to one-point closure models, which are based on time averaging of the Navier-Stokes equations, LES is based on the solution of the spatially filtered equations, see e. g. [6]. A subgrid scale model is then used to model the filtered turbulent scales. Here, LES of a internal flows in bended pipes, pipe junctions and jet apparatus and in free turbulent jets including the mixing of different species are presented. Results of validation test cases and details of the algorithm applied have been reported in [18].

Results of the flow in a 90° bend for a Reynolds number of 5000 based on the mean velocity and the pipe diameter, a Dean number of 1800, and a Mach number of 0.1 is shown in Fig. 4-5. Flow separation occurs at the inner side behind the bend and also on the outer side in a small region near the beginning of the bend. The comparison of the time averaged results with experimental data shows good agreement in the radial cross sections. In both the numerical and experimental data, an oscillation of the Dean vortices was found. More details of this problem will be published in a separate paper.

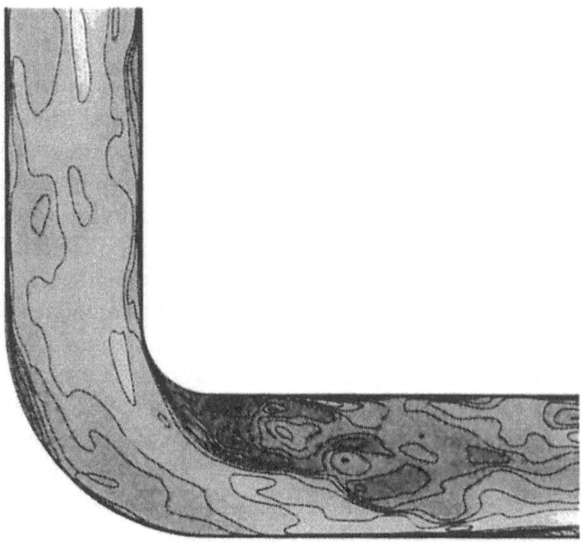

Figure4. Instantaneous flow field in a axial section of a turbulent flow in a 90° bend. Gray scales and lines of constant Mach number for a Reynolds number of 5000 and a Dean number of 1800.

A careful formulation of approximate in- and outflow boundary conditions is important in order to avoid unphysical results. For the inflow boundary for the bend, a solution is extracted for each time step from a simulation of a turbulent pipe flow with periodic conditions in streamwise direction. The simulation for

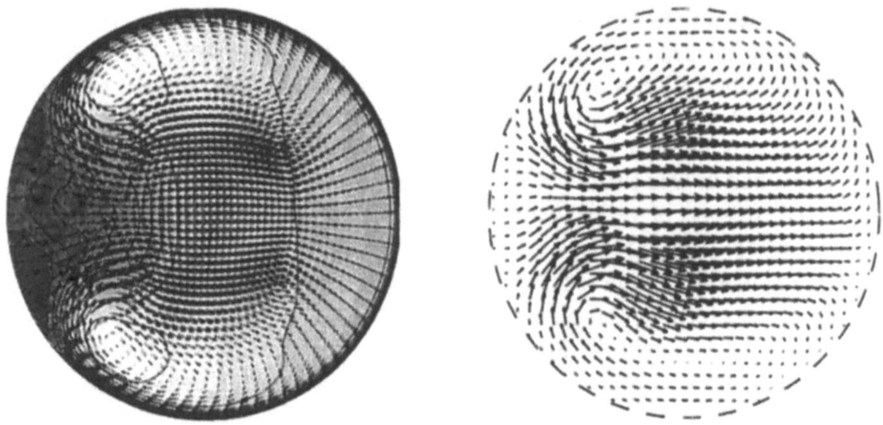

Figure5. Time averaged velocity vectors of a turbulent flow in a radial cross section at the end of a 90° pipe bend. Comparison of the numerical solution (left) with experimental data (right) for a Reynolds number of 5000 and a Dean number of 1800.

the pipe and the bend are conducted in parallel, so that physical meaningful instantaneous values of the mass flow can be provided at the inflow boundary.

The flow in a T-junction of two pipes with a diameter ratio of 0.425 was simulated for laminar inflow conditions. This flow exhibits a large number of different flow states, dependent on the Reynolds number of the flow and the ratio of the mass fluxes in the main and secondary pipe. Two different cases are shown in Fig. 6-7. In the case of the lower Reynolds number the volume rates in both pipes is comparable, so that the velocity in the secondary pipe is relatively high and a jet like flow issues into the crossflow of the main pipe and then hits the opposite pipe wall. A complicated vortex system is generated with a horseshoe vortex near the location where the jet hits the opposite wall and a periodic shedding of vortices behind the jet. For the higher Reynolds number the volume rate in the secondary pipe is relatively small, so that the jet is redirected into the flow direction of the main pipe. In this case a pulsating flow establishes, which is better visible in experimental flow visualization.

As an example for free turbulent flows, a round jet evolving into the same fluid at rest is simulated. The instantaneous surface of constant vorticity shows the large scale structures of the jet in Fig. 8. The decay of the centerline velocity and the velocity profiles are shown in comparison with experimental data in Fig. 9. Good agreement with experimental data can be observed.

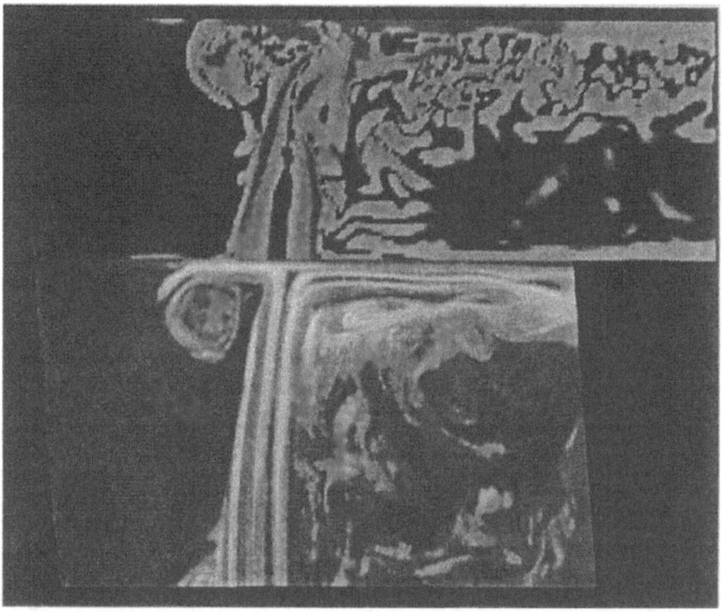

Figure6. Flow field in a 90° T-junction. $\dot{Q}_0/\dot{Q}_1$=5/6, *Re*=680. Comparison of the numerical solution (top) with experimental flow visualization (bottom).

Figure7. Flow field in a 90° T-junction. $\dot{Q}_0/\dot{Q}_1$=2, *Re*=3630. Comparison of the numerical solution (top) with experimental flow visualization (bottom).

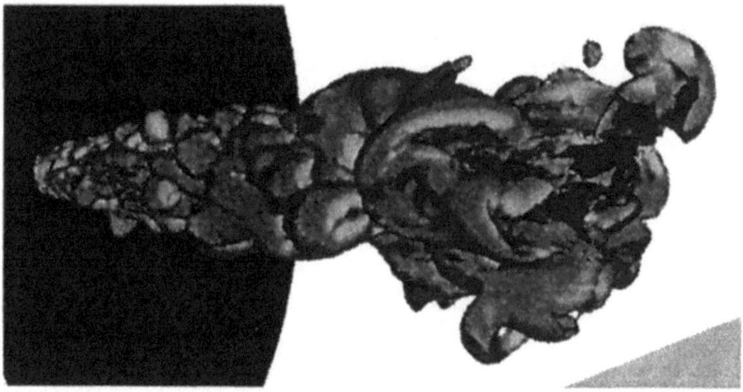

Figure8. Instantaneous flow field of a turbulent round jet. Surface of constant vorticity.

The mixing of gases is an important technical problem. It is investigated here with an algorithm, which solves additional conservation equations for the partial densities of an arbitrary number of species, details see [17]. A LES of the flow in a jet apparatus, in which an air jet from a fully developed pipe flow sucks Argon from stagnation conditions into the apparatus. The Reynolds number of the pipe flow is 5000, the Mach number is 0.1. The mixing processes of the two fluids are shown in Fig. 10 where the instantaneous flow field of the concentration of N_2 is shown. Detailed statistical data is currently obtained for this case.

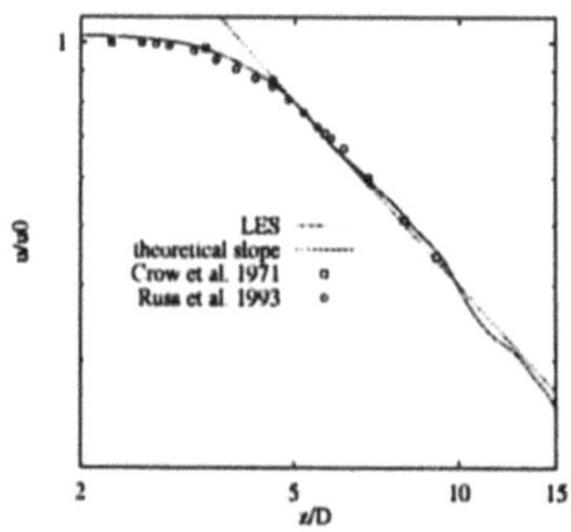

Figure9. Time averaged centerline velocity of an air-in-air jet, LES at $Re = 20000$ in comparison to experimental results, [5, 19]

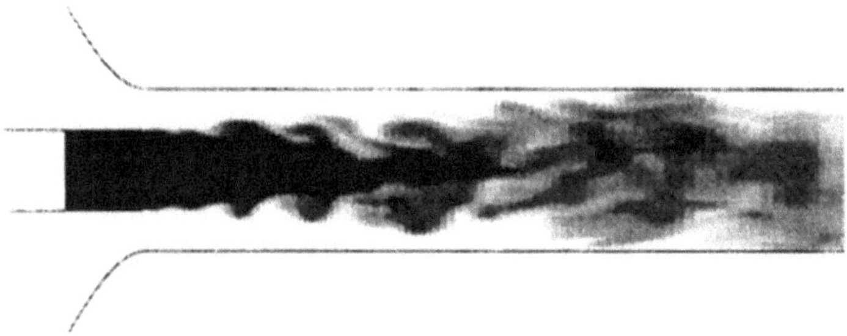

Figure10. Instantaneous flow field in a jet apparatus. Mixing of air with argon in an axial section.

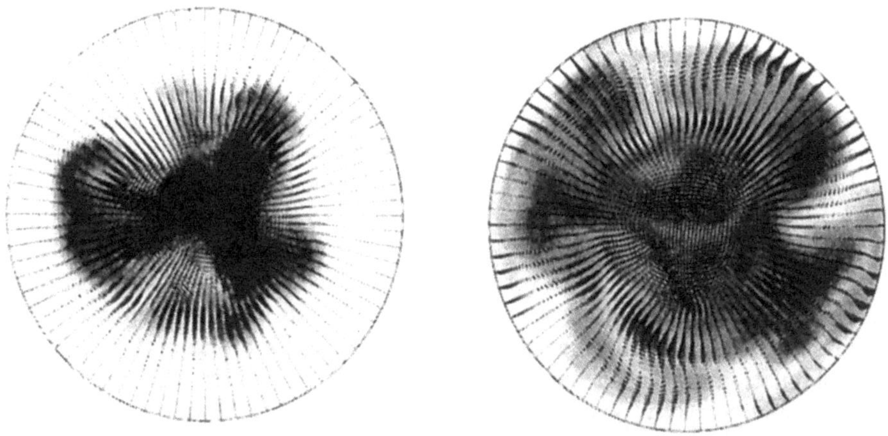

Figure11. Instantaneous flow field in a jet apparatus. Concentration of argon with velocity vectors in two radial cross sections close to the exit of the inner pipe (left) and further downstream (right).

3.3 Simulation of a Piston Engine Flow

Understanding the unsteady, three-dimensional, compressible flow field structure in piston engines during the intake and compression stroke is crucial for the development of new engine designs with high performance and low emission values. The flow field determines the flame propagation rate in homogeneous charge spark-ignition engines and the fuel-air mixing and burning rates in Diesel engines, [9]. It has a major impact on the emission values and the breathing capacity of the engine and therefore also on the maximum available power.

Separation of the flow can occur at the valve seat and valve head depending on the seat angle, the rounding of the seat corners, the seat width and the swirl generated in the intake port. An investigation of the above mentioned phenomena with a numerical solution of the Navier-Stokes equations requires the resolution of all important details of the intake and cylinder head geometry. This

Figure14. Bubble type vortex breakdown in a circular divergent pipe: Surface of constant pressure.

The flow in a piston engine was simulated on a moving grid with a block structured mesh showing qualitative agreement with experimental data. The solution was used to analyse the vortex topology of the flow field. It could be shown, that starting during the intake stroke two ring-like vortices are generated, which later merge with other secondary vortices to a single vortex structure. This solution will be further developed with the aim to carry out a LES with a resolution of the larger turbulent scales.

The problem of vortex breakdown was successfully simulated in a circular divergent pipe. A steady bubble type breakdown could be predicted with close agreement to experimental data.

Acknowledgments

The work reviewed here was sponsored by the German Research Association (DFG). The authors acknowledge contributions of A. Abdellfattah, A. Henze, M. Kropp, Th. Rister, F. Rütten, C. Schulz, A. Schvorak and M. Weimer for the different applications presented.

References

1. A. Abdelfattah. *Numerische Simulation von Strömungen in 2- und 4-Ventil Motoren.* Dissertation, Aerodynamisches Institut, RWTH Aachen, 1997.
2. W. Althaus, Ch. Brücker, and M. Weimer. Breakdown of slender vortices. In S.I.Green, editor, *Fluid Vortices.* Kluwer Academic Publishing, 1994.
3. M. Breuer and D. Hänel. A dual time-stepping method for 3-d, viscous, incompressible vortex flows. *Computers Fluids*, 22(4/5):467–484, 1993.

4. A.J. Chorin. A Numerical Method for Solving Incompressible Viscous Flow. *Journal of Computational Physics*, 2:12–26, 1967.
5. S. Crow and F. Champagne. Orderly structure in jet turbulence. *Journal of Fluid Mechanics*, 48:547–591, 1971.
6. G. Erlebacher, M. Y. Hussaini, C. G. Speziale, and T. A. Zang. Toward the Large-Eddy Simulation of compressible turbulent flows. *Journal of Fluid Mechanics*, 238:155–185, 1992.
7. A. Henze. *Integration der Navier-Stokes-Gleichungen für Hyperschallströmungen mit Realgaseffekten*. PhD thesis, RWTH Aachen, 1995.
8. A. Henze, E.M. Houtman, M. Jacobs, and V.N. Vetlutsky. Comparison between experimental and numerical heat flux data for supersonic flow around ELAC-1. *Z. Flugwiss. Weltraumforsch.*, 20, 1996.
9. J. B. Heywood. *Combustion Engine Fundamentals*. McGraw-Hill, 1988.
10. A. Jameson, W. Schmidt, and E. Turkel. Numerical Solution of the Euler equations by finite volume methods using Runge-Kutta time stepping schemes. *AIAA Paper*, No. 81-1259, 1981.
11. E. Krause. The Solution to the Problem of Vortex Breakdown. In *Lecture Notes in Physics* vol. 371, pages 35–50. Invited Lecture at the 12th International Conference on Numerical Methods in Fluid Dynamics, Oxford, England, July 9-13, Springer Verlag, 1990.
12. M. Kropp. *Reagierende Über- und Hyperschallumströmung eines Raumtransportsystems mit Außenverbrennung*. Dissertation, Aerodynamisches Institut, RWTH Aachen, 1997.
13. B. P. Leonard. A stable and accurate convective modelling procedure based on quadratic upstream interpolation. *Computer Methods in Applied Mechanics and Engineering*, 19:59, 1979.
14. S. Liu, C. Meneveau, and J. Katz. On the Properties of Similarity Subgrid-scale Models as Deduced from Measurements in a Turbulent Jet. *Journal of Fluid Mechanics*, 275:83–119, 1994.
15. M. Meinke, A. Abdellfattah, and E. Krause. Simulation of Piston Engine Flows in Realistic Geometries. In J.-J. Chattot P. Kutler, J. Flores, editor, *15th International Conference on Numerical Methods in Fluid Dynamics*, pages 195–200. Springer, June 1996.
16. M. Meinke, A. Abdellfattah, and E. Krause. Simulation and Flow Field Analysis of Piston Engine Flows. In K. Kuwahara, editor, *5th Nobeyama Workshop on High-Performance Computing in Fluid Dynamics, May 13-15, 1997, Nobeyama, Japan.* to be publish in Fluid Dynamics Research, Elsevier, 1997.
17. M. Meinke, J. Hofhaus, and A. Abdellfattah. Vortex Ring Interaction. In E. Krause, editor, *IUTAM Symposium on Dynamics of Slender Vortices, September 1997, Aachen, Germany.* to be publish in Kluwer, 1997.
18. M. Meinke, C. Schulz, and Th. Rister. Large Simulation of a Spatially Developing Jet. In *Notes on Numerical Fluid Mechanics: Computation and Visualization of three-dimensional vortical and turbulent flows.* Vieweg Verlag, to be published, 1997.
19. S. Russ and P. J. Stykowski. Turbulent structure and entrainment in heated jets: The effect of initial conditions. *Physics of Fluids*, A 5(12):3216–3225, December 1993.
20. J. L. Shinn, H. C. Yee, and K. Uenishi. Extension of a Semi-Implicit Shock-Capturing Algorithm for 3-d Fully Coupled, Chemically Reacting Flows in Generalized Coordinates. *AIAA Paper*, No. 87-1577, June 1987.

21. A. Stromberg, A. Henze, W. Limberg, and E. Krause. Investigation of vortex structures on delta wings. *Z. Flugwiss. Weltraumforsch.*, 20, 1996.
22. P. D. Thomas and C. K. Lombard. Geometric Conservation Law and Its Application to Flow Computations on Moving Grids. *AIAA Journal*, 17(10):1030–1037, October 1979.
23. M. Weimer. *Aufplatzen freier Wirbel und drallbehafteter Rohrströmungen*. Dissertation, Aerodynamisches Institut, RWTH Aachen, 1997.

The Decay of Three-dimensional Deviations from the Fully Developed State in Laminar Pipe Flow

K. Gersten, M. Klika
Institut für Thermo- und Fluiddynamik
Ruhr-Universität Bochum, Germany

The basic equations for the flows in circular pipes at high Reynolds numbers are derived. At a certain cross-section a three-dimensional velocity field is given which is only slightly different from the fully developed state. The decay of the deviations is investigated by linearising the basic equations leading to an eigenvalue problem. Eigenvalues and eigenfunctions are given. The decaying flow consists of various flow structures such as ring vortices as well as single, double and multiple streamwise vortices. Each of these flow structures have different decay behaviour. The swirl (single streamwise vortex) has the lowest decay. The decaying is described by means of characteristic parameters, which correspond with the various flow structures.

1. Introduction

The flows in straight pipes are often not fully developed. Deviations from the fully developed state may occur due to flows in bends, valves or other pipe fittings. Even swirl can be found downstream of two bends in perpendicular planes. The decay of the deviations from the fully developed condition is of great practical importance. For example, these deviations can cause significant errors in the measurements of the mass flow rate. It is therefore important to know what length of straight pipe is required for a given upstream disturbed flow to reduce to the specified acceptable deviation from the fully developed flow condition. It is also worth to know the additional pressure drop due to the disturbances.

2. Basic Equations

The laminar incompressible steady flow in a straight pipe is considered, see Fig.1. The basic equations in vector form are as follows, see Schlichting, Gersten (1997):

continuity equation

$$\text{div } \vec{w}^* = 0, \tag{1}$$

momentum equation

$$\text{grad } \left(\frac{1}{2}\vec{w}^{*2}\right) - \vec{w}^* \times \vec{\omega}^* = -\frac{1}{\rho^*} \text{ grad } p^* + \nu^* \Delta \vec{w}^*, \tag{2}$$

where

$$\vec{\omega}^* = \text{ curl } \vec{w}^* \tag{3}$$

is the vorticity. The curl of Eq.(2) leads to the vorticity transport equation

$$\text{curl } (\vec{w}^* \times \vec{\omega}^*) + \nu^* \Delta \vec{\omega}^* = 0. \tag{4}$$

Dimensional quantities are marked by a star. A cylindrical coordinate system is used, see Fig.1.

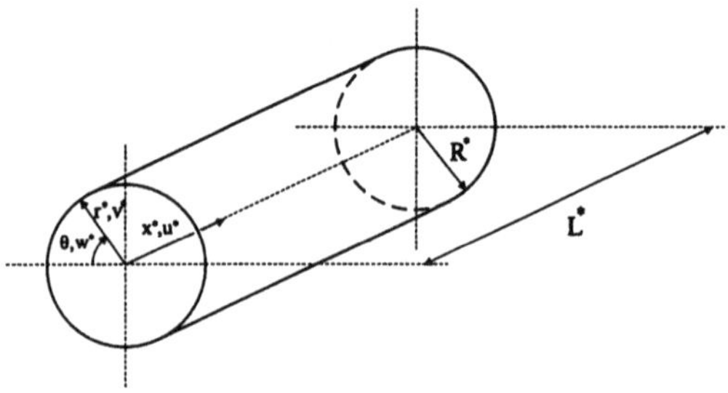

Figure 1: Geometry and Coordinate System

The following dimensionless coordinates are introduced:

$$x = \frac{x^*}{L^*} \quad , \quad r = \frac{r^*}{R^*} \quad , \tag{5}$$

where R^* is the pipe radius and L^* a length in axial direction, characteristic for the decaying of the flow disturbances considered. Hence the ratio

$$\epsilon_1 = \frac{R^*}{L^*} \tag{6}$$

is supposed to be a small number. The axial velocity component is based on the mean velocity

$$u_m^* = \frac{1}{\pi R^{*2}} \int_0^{2\pi} \int_0^{R^*} u^*(x^*, r^*, \theta) r^* \, dr^* d\theta. \tag{7}$$

To avoid a degeneration of the continuity equation the velocity components v^* and w^* are based on $\epsilon_1 u_m^*$, in other words they are small compared to u_m^*. Hence, the following dimensionless quantities are used:

$$u = \frac{u^*}{u_m^*} \quad , \quad v = \frac{v^*}{\epsilon_1 u_m^*} \quad , \quad w = \frac{w^*}{\epsilon_1 u_m^*} \quad , \quad p = \frac{p^* - p_0^*}{\rho^* u_m^{*2}} \quad ,$$

$$\omega_x = \frac{1}{\epsilon_1} \frac{\omega_x^* R^*}{u_m^*} = \frac{R^*}{\epsilon_1 u_m^* r^*} \left[\frac{\partial (r^* w^*)}{\partial r} - \frac{\partial v^*}{\partial \theta} \right]. \tag{8}$$

The Reynolds number is assumed to be large:

$$Re = \frac{2R^* u_m^*}{\nu^*} = \frac{1}{\epsilon_2}. \tag{9}$$

By using these dimensionless quantities the basic equations in the cylindrical coordinate system follow:

$$\frac{\partial u}{\partial x} + \frac{1}{r} \frac{\partial (rv)}{\partial r} + \frac{1}{r} \frac{\partial w}{\partial \theta} = 0 \, , \tag{10}$$

$$u \frac{\partial u}{\partial x} + v \frac{\partial u}{\partial r} + \frac{w}{r} \frac{\partial u}{\partial \theta} = -\frac{\partial p}{\partial x} + 2 \frac{\epsilon_2}{\epsilon_1} \left[\epsilon_1^2 \frac{\partial^2 u}{\partial x^2} + \frac{1}{r} \frac{\partial}{\partial r} \left(r \frac{\partial u}{\partial r} \right) + \frac{1}{r^2} \frac{\partial^2 u}{\partial \theta^2} \right] , \tag{11}$$

$$\epsilon_1^2 \left[u \frac{\partial v}{\partial x} + v \frac{\partial v}{\partial r} + \frac{w}{r} \frac{\partial v}{\partial \theta} - \frac{w^2}{r} \right] = -\frac{\partial p}{\partial r} + 2\epsilon_1 \epsilon_2 \left[\epsilon_1^2 \frac{\partial^2 v}{\partial x^2} + \frac{d}{dr} \left(\frac{v}{r} \right) + \frac{\partial^2 v}{\partial r^2} + \frac{1}{r^2} \frac{\partial^2 v}{\partial \theta^2} - \frac{2}{r^2} \frac{\partial w}{\partial \theta} \right] , \tag{12}$$

$$\epsilon_1^2 \left[u \frac{\partial w}{\partial x} + v \frac{\partial w}{\partial r} + \frac{vw}{r} + \frac{w}{r} \frac{\partial w}{\partial \theta} \right] = -\frac{1}{r} \frac{\partial p}{\partial \theta} + 2\epsilon_1 \epsilon_2 \left[\epsilon_1^2 \frac{\partial^2 w}{\partial x^2} + \frac{d}{dr} \left(\frac{w}{r} \right) + \frac{\partial^2 w}{\partial r^2} + \frac{1}{r^2} \frac{\partial^2 w}{\partial \theta^2} + \frac{2}{r^2} \frac{\partial v}{\partial \theta} \right] . \tag{13}$$

To avoid the degeneration of the momentum equation the two limits $\epsilon_1 \to 0$ and $\epsilon_2 \to 0$ have to be taken such that

$$\epsilon_1 = \epsilon_2 = \epsilon. \tag{14}$$

	i	1	2	3	4	5
$k=0$	Λ_{wi}	21.38	74.78	160.2	277.5	426.9
swirl	$W_{0i}'''(0)$	-16.07	-55.93	-119.3	-205.6	-314.1
$W_{0i}'(0)=1$	$W_{0i}'(1)$	-0.502	0.335	-0.261	0.218	-0.189
$k=0$	Λ_{0i}	32.08	95.68	191.4	319.1	478.8
no swirl	$U_{0i}''(0)$	-13.10	-53.38	-85.68	-160.9	-207.6
$U_{0i}(0)=1$	$U_{0i}'(1)$	1.453	-2.678	2.642	-3.489	3.449
$k=1$	Λ_{1i}	44.83	96.65	197.8	321.0	519.1
$V_{1i}(0)=1$	$U_{1i}'(0)$	0.393	0.002	0.013	0.001	0.071
	$V_{1i}''(0)$	-1.711	-12.26	-22.49	-40.37	-62.00
	$U_{1i}'(1)$	0.135	-0.099	0.035	-0.042	0.026
	$W_{1i}'(1)$	-3.916	3.308	-3.559	4.068	-4.550
$k=2$	Λ_{2i}	52.70	131.3	242.2	382.3	592.9
$W_{2i}'(0)=1$	$U_{2i}''(0)$	0.237	-0.391	0.106	-0.112	0.038
	$W_{2i}'''(0)$	-8.519	-10.19	-43.02	-50.84	-110.5
	$U_{2i}'(1)$	-0.023	0.032	-0.007	0.009	-0.003
	$W_{2i}'(1)$	0.571	-0.483	0.459	-0.367	0.401

Table 1: Eigenvalues and essential quantities of eigenfunctions

for $k = 2$ can be seen in Fig.5. The two counter-rotating axial vortices can clearly be recognized.

The indices w and k of the eigenvalues correspond to various flow structures which have different decay behaviour. The swirl of the single streamwise vortex (index w) has the lowest decay. The decay is higher for the ring vortices and even higher for two counter-rotating streamwise vortices ($k = 1$, Fig.5a), for four streamwise vortices ($k = 2$, Fig.5b), and so on.

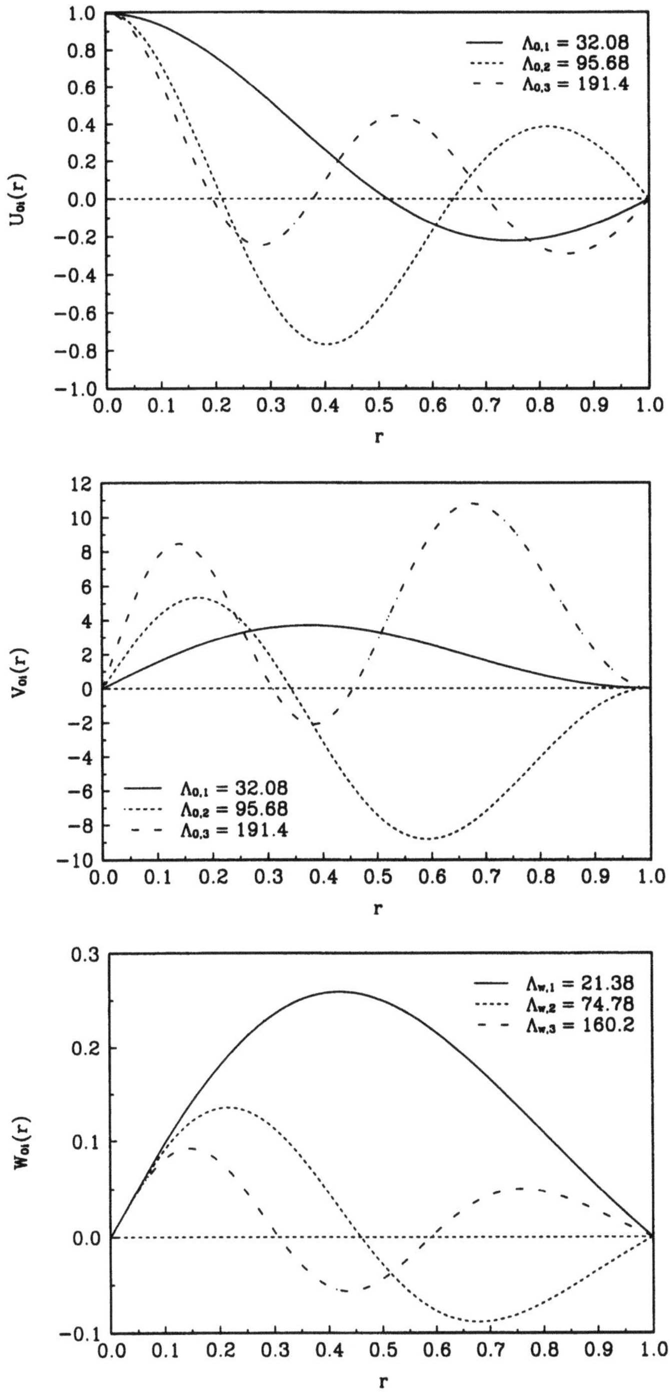

Figure 2: Eigensolutions for $k = 0$

a

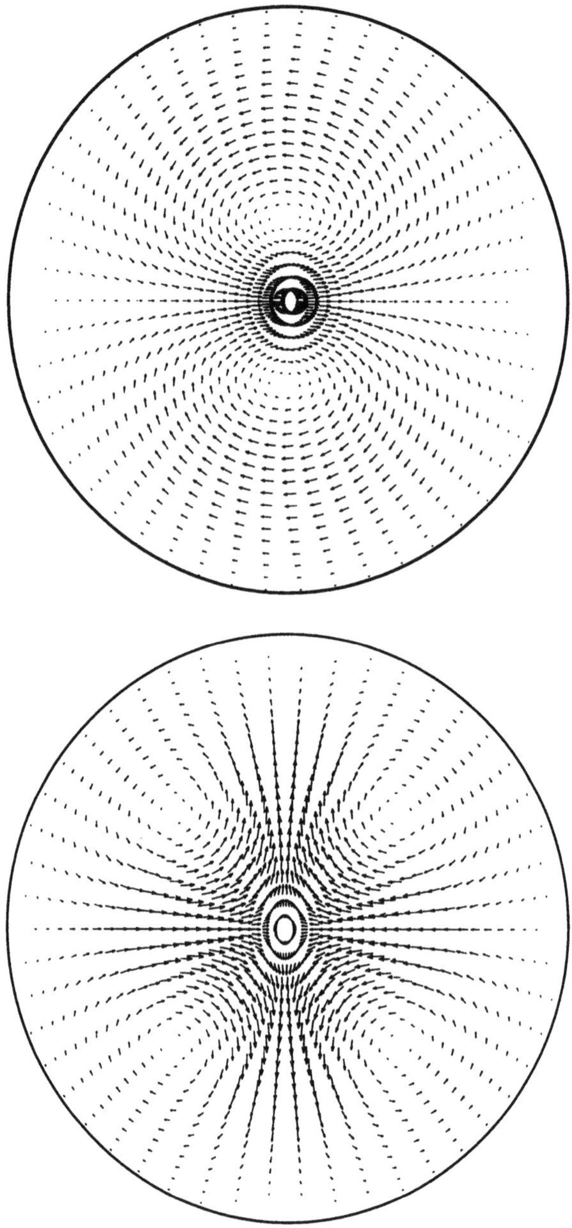

b

Figure 5. Vector plot of the secondary flow (V_{ki}, W_{ki}) in section $x = 0$.

a) $k = 1$, first eigensolution

b) $k = 2$, first eigensolution

5 Characteristic Parameters

For flow metering and for other technical applications it is common to use global parameters to characterize the flow in a pipe, see Morrison et al. (1992,1995) and Steenbergen (1995). Here the following parameters are suggested:

$$K_m = \frac{1}{\pi}\int_0^{2\pi}\int_0^1 u^2 r \cos m\theta \, dr d\theta \qquad m = 0, 1, 2, \ldots \tag{29}$$

$$S_m = 2\int_0^{2\pi}\int_0^1 u \frac{w^*}{u_m^*} r^2 \cos m\theta \, dr d\theta \qquad m = 0, 1, 2, \ldots \quad . \tag{30}$$

Similar parameters can be defined by using $\sin m\theta$ instead of $\cos m\theta$ in Eqs.(29) and (30). The parameter K_0, also called *momentum correction factor*, see Gersten (1983), describes the decay of the ring vortex:

$$K_0 = 1.333 + \sum_{i=1} A_{0i} C_{Ki} e^{-\Lambda_{0i} x} \tag{31}$$

where

$$C_{Ki} = 4 \int_0^1 U_\infty(r) U_{0i}(r) r \, dr \ . \tag{32}$$

The decay of the single streamwise vortex is given by S_0, also called *swirl number*, see Steenbergen (1995):

$$S_0 = \epsilon_1 \sum_{i=1} B_i C_{Si} e^{-\Lambda_{wi} x} \tag{33}$$

where

$$C_{Si} = 4\pi \int_0^1 U_\infty(r) W_{0i}(r) r^2 \, dr. \tag{34}$$

The swirl number is directly related with the circumferential component of the wall shear stress by the global balance equation of angular momentum:

$$\frac{d}{dx}\left(\frac{S_0}{\epsilon_1}\right) = 8\pi \left(\frac{dw}{dr}\right)_{r=1} . \tag{35}$$

In the same way S_1 determines the decay of the two counter–rotating streamwise vortices, whereas S_2 describes the decay of the four streamwise vortices and so on. The more global parameters are used the better the three–dimensional velocity distributions in the pipe flow can be described and the decaying behaviour be determined.

computer codes which have been applied for turbomachinery blade analysis in industrial use Dawes' *BTOB3D* [4] produces results on simple meshes with a simple mixing length model which offer some global information about the effect of design parameters on loss production. However, for appropriate loss predictions and computation of the behaviour of the boundary layers, more refined meshes have to be used, and the pretension of the user determines the effort required to achieve the desired level of resolution of details of the shear layers or secondary flow domains. As in case of empirical loss models Navier-Stokes solvers need careful calibration, too, because improper cell sizes in the near wall boundary layer region may have a strong impact on the numerical results and can give rise to misleading interpretation of the results. Based on Navier-Stokes solvers developed by the author [5,6] which supplement the design systems for turbomachinery bladings these aspects will be high-lighted and calibration criteria discussed.

2 Quasi-3D viscous flow analysis

A 2D viscous solver by the author [5] is included in a quasi-3D design approach following Wu's general approach [1] and is used in the blade-to-blade surfaces which are defined by the streamlines of a axisymmetric through-flow analysis in the meridional $S_{2m}-$ streamsheet. The numerical scheme contains explicit Runge-Kutta time stepping and finite difference space discretization. The purpose of the viscous flow analysis is evaluation of the aerodynamic properties for the corresponding airfoil sections at the individual radial stacking locations. Through varying stream sheet thickness and curved streamlines in the $S_{2m}-$ surfaces the blade-to-blade analysis can resolve 3D-effects to some extent. If such a solver is integrated into a design system the computational meshes must be generated automatically [5] to avoid user's ambiguity. As an example the mesh arrangement containing 127×49 grid nodes is shown for the flow analysis of the VKI- gas turbine profile section [7]. Essential features of the meshes generated with an algebraic preprocessor developed by the author are quasi-orthogonality of the mesh near the surfaces to enable more accurate predictions of boundary layer parameters and the resolution of flow details around leading and trailing edges. For different exit Mach numbers the isentropic Mach number distributions are shown in Fig. 2 with both an algebraic Baldwin-Lomax model and a k- ϵ model for the turbulent flows. The algebraic model is combined with a transition model of Abu-Ghannam and Shaw [8]. The more elaborate k- ϵ model is extended towards the wall with low-Reynolds number terms by Chien (1982). The latter model is very sensitive with regard to the cell sizes, and a very high resolution of the near wall boundary layer must be achieved to obtain results which can be regarded almost mesh-independent. For 2D- applications the k- ϵ model offers only marginal improvements which do not justify its application. A blow-up figure of the flow near the trailing edges for both cases, Fig. 3, shows details of the shock-boundary layer interaction and does not indicate flow separation in either case which is an important information for the blade designer.

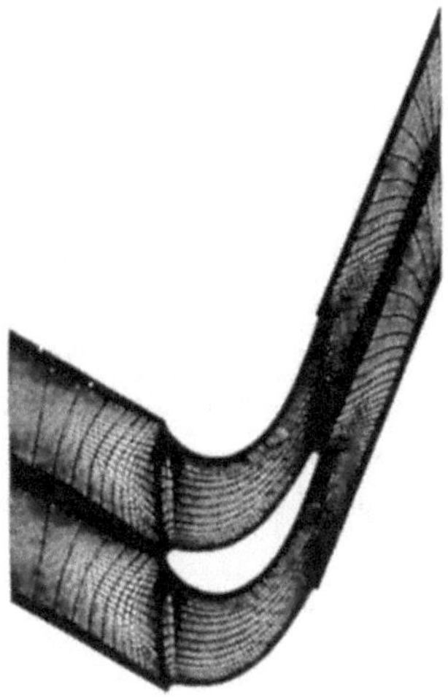

Fig. 1: Mesh for the VKI - profile [7], 127 × 49 nodes.

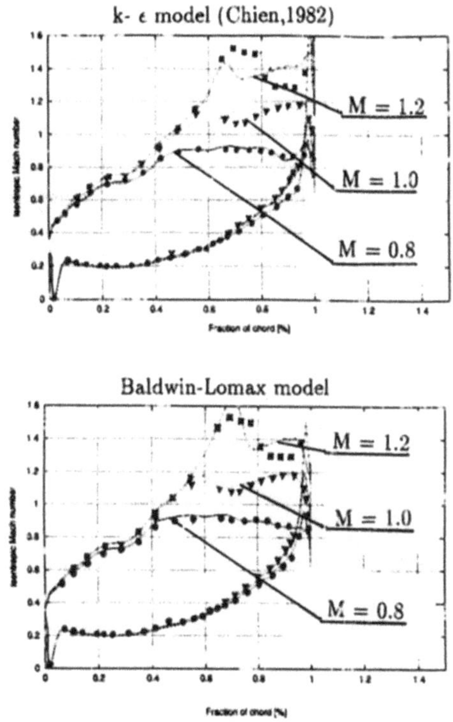

Fig. 2: Isentropic Mach number contours along the blade surfaces; experiments ref. [7].
Comparison of turbulence models.
M: Exit Mach number

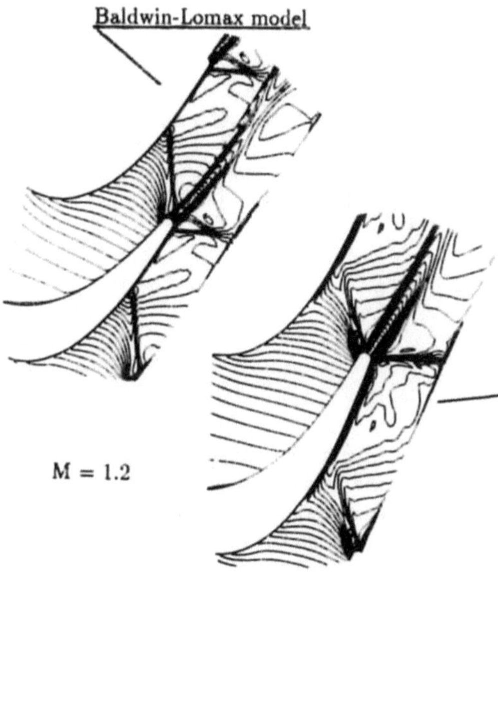

Fig. 3: Mach number distribution near the trailing edge of the VKI-profile

Another example with the quasi-3D solver is the flow around a CDA [1] -airfoil, Fig. 4 [9]. For design and off-design operating points the agreement of measured and predicted isentropic Mach number distributions along the surfaces, Fig. 5, is good if the cheaper Baldwin-Lomax model is applied. An important property of the solver is the prediction of boundary layer parameters. The skin friction coefficients indicate transition of the boundary layer well downstream of the leading edge of the airfoil which was intended for this design at the linear wind tunnel test conditions with an inflow turbulence level of about 0.3%, [9]. Fig. 6 shows comparison of measured and predicted loss bucket (= loss vs. incidence) graphs for various Mach number ranges and demonstrates that the current solver predicts the operational range of the airfoil section properly. In order to gain confidence in the applicability of the code a great number of such test cases has to be run.

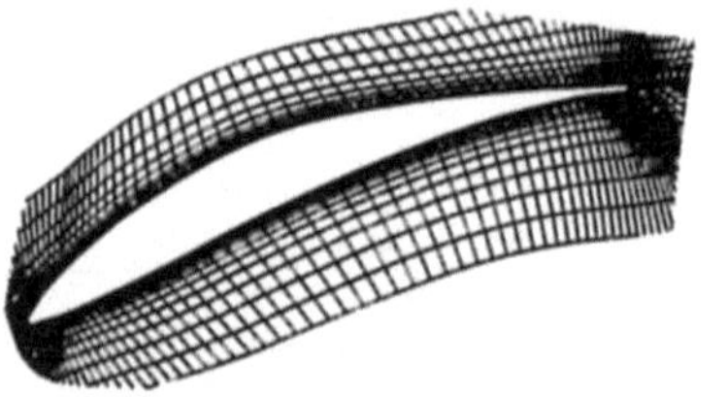

Fig. 4: Detail of the mesh around a CDA - profile; 125 × 41 nodes.

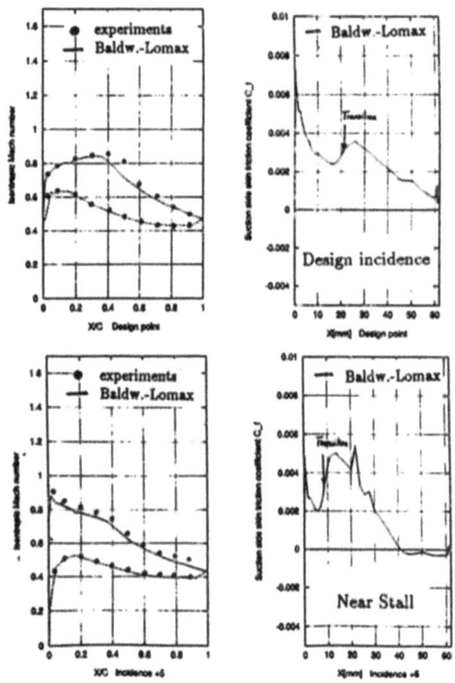

Fig. 6: Comparison of measured [9] and predicted loss characteristics of a CDA-profile ([5])

Fig. 5: Comparison of measured [9] and predicted isentropic Mach number contours and skin friction coefficients c_f for design and near stall incidence ([5])

---

[1]CDA= Controlled Diffusion Airfoil

3 Fully 3D viscous flow analysis

Similar to the quasi-3D code explicit Runge-Kutta time marching with residual averaging is applied in a 3D-Navier Stokes code of the author [6]. In order to accommodate high aspect ratio cells caused by the fine resolution of the boundary layer, parameters of the residual averaging matrices and the artificial viscosity terms are carefully weighted using the eigenvalues of the governing equations for each individual spatial direction [10]. During the code calibration process it became obvious that only the more expensive two-equation model with low-Reynolds number terms (Chien, 1982) provided results of the desired level of accuracy. The boundary layer is fully resolved including the extremely thin laminar sublayer. For an axial compressor flow analysis the velocity profiles are plotted in Fig. 7 along different wetted surfaces, and it can be seen that the value for the characteristic boundary layer coordinate y^+ is of the order of unity for the first grid points away from the wall. Fig. 8 shows the computed and measured total pressure ratios and adiabatic efficiencies of the well known Rotor 67 by *NASA* tested by Stratizar et. al. [11]. The numerical analysis was run with a mesh containing 137 grid points in streamwise, 49 in pitchwise and 59 in spanwise direction. With full resolution of boundary layers and details around leading and trailing edges a high level of accuracy can be achieved with regard to important global design parameters of the row as this example demonstrates.

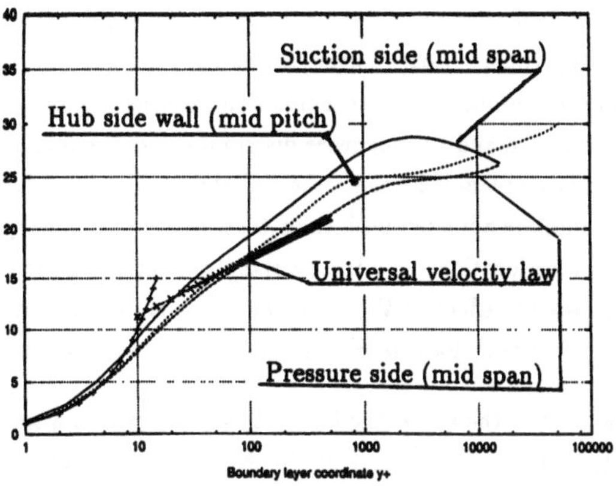

Definitions:

$$y^+ = \frac{\rho u_\tau y}{\mu}$$

$$u^+ = \frac{u}{u_\tau}$$

$$u_\tau = \sqrt{\frac{\tau}{\rho}}$$

u : flow velocity
τ: wall shear stress
y: normal wall distance
ρ: density
μ: dynamic viscosity

Fig. 7: Comparison of computed velocity profiles with the universal logarithmic velocity profile for unseparated flow in a transonic axial compressor rotor at different positions

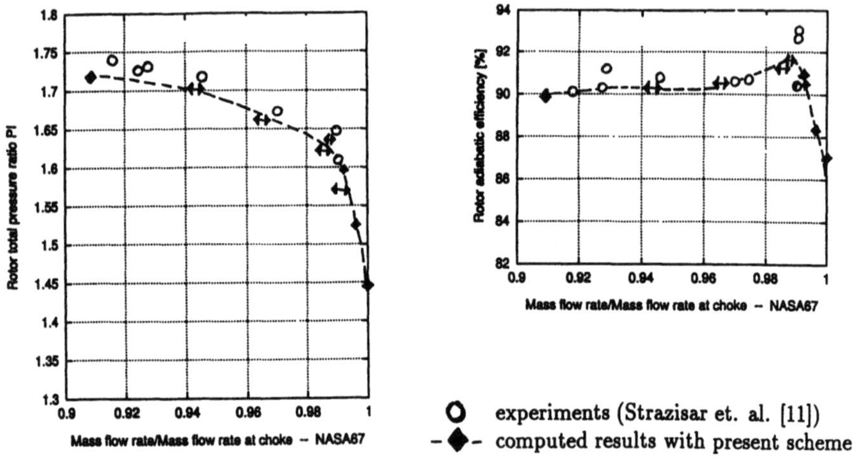

Fig. 8: Comparison of predicted and measured total pressure ratios and adiabatic efficiencies for NASA-Rotor 67 [11]
(Design pressure ratio *1.63*, design tip Mach number $M = 1.32$)

Moreover, for analysis of blade rows in multi-stage turbomachines the effect of the neighboring blade rows on the flow field solution has to be included. To achieve this the flow field computations with the 3D Navier-Stokes solver can be coupled, and the boundary conditions are updated using the results of the other blade row during the flow computation. The meshes can be overlapped and the spanwise distributions of the individual row analysis are compared at a mixing plane between rotor and stator. The boundary conditions at the inlet and exit planes are adjusted until the pitchwise averaged flow quantities are in agreement at all spanwise locations. At the overlapping regions of the extended meshes and the vaned passages the effect of the next blade row is included in the flow equation using a simple axisymmetric blockage model and potential body forces.

With the multi-row arrangement of the solver the flow field was studied in a single stage axial compressor investigated by Dunker et. al. [12]. In these experiments two different stators were run, namely one with 60 conventional *NACA*- airfoils, and another with 31 CDA-vanes. Comparisons of measured and predicted total pressure profiles and losses at the stator discharge planes, Fig. 9, confirm the characteristic slopes inherent with root separation which was indeed observed for the redesigned CDA-stator. The corresponding streamline patterns near the wetted surfaces are shown in Fig. 10.

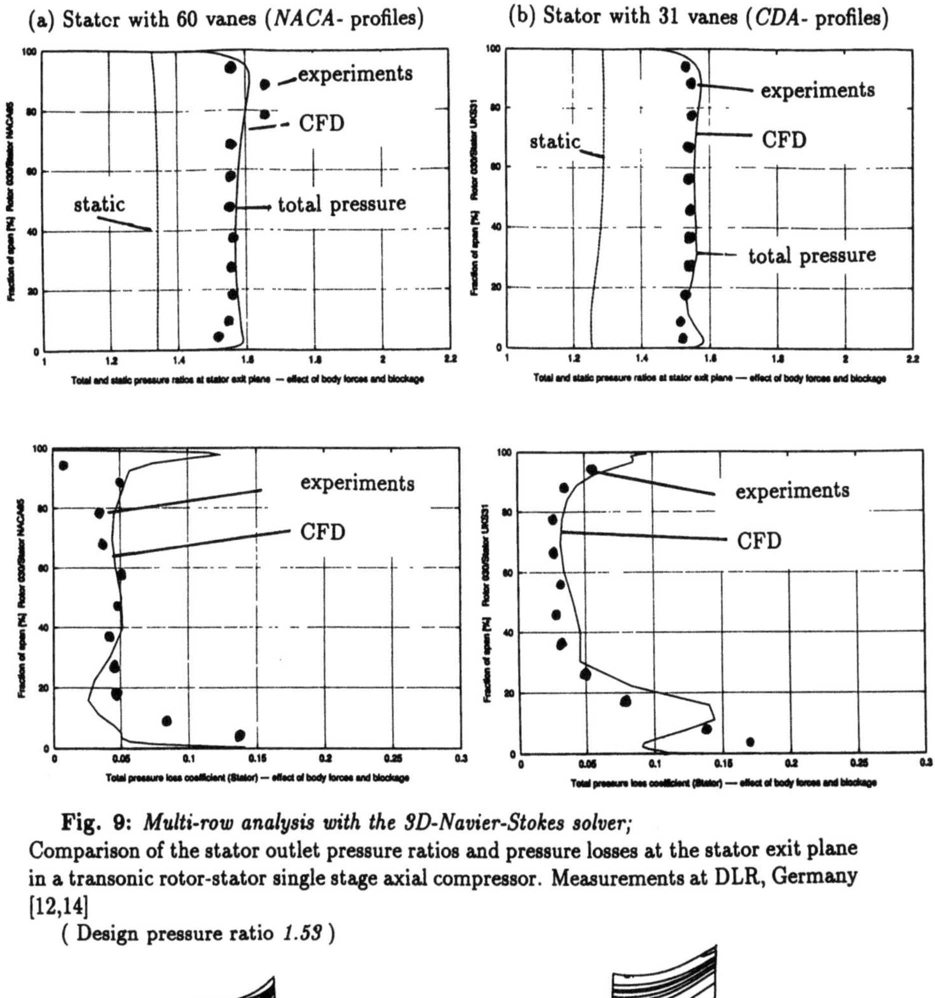

Fig. 9: *Multi-row analysis with the 3D-Navier-Stokes solver;*
Comparison of the stator outlet pressure ratios and pressure losses at the stator exit plane in a transonic rotor-stator single stage axial compressor. Measurements at DLR, Germany [12,14]
(Design pressure ratio *1.53*)

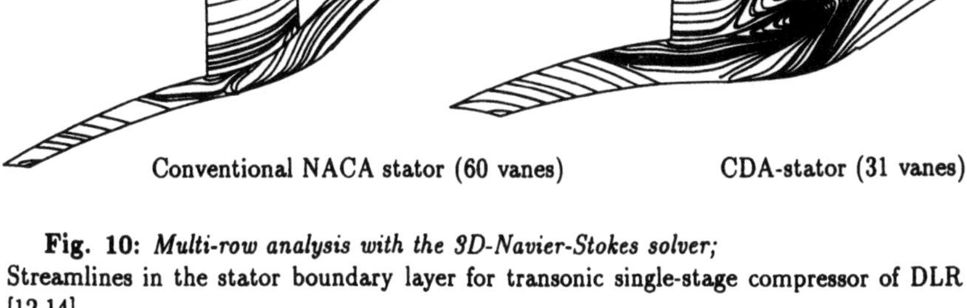

Conventional NACA stator (60 vanes) CDA-stator (31 vanes)

Fig. 10: *Multi-row analysis with the 3D-Navier-Stokes solver;*
Streamlines in the stator boundary layer for transonic single-stage compressor of DLR [12,14]

References

[1] Wu, C.-H.; A General Theory of Three- Dimensional Flow in Subsonic Turbomachines of Radial- , Axial- , and Mixed Flow Types. *NACA TN-D* 2604, 1952

[2] Denton, J.D.; A Time Marching Method for Two- and Three- Dimensional Blade to Blade Flows. *ARC R&M* 3775, 1975

[3] Denton, J.D.; The Calculation of Three-Dimensional Viscous Flow Through Multistage Turbomachines, *ASME J. of Turbomachinery*, Vol. 114, pp. 18-26, 1992

[4] Dawes, W.N.; Development of a 3D Navier-Stokes Solver for Application to All Types of Turbomachinery. *ASME* 88-GT-70, 1988

[5] Wiedermann, A., Kumar, S.; Navier-Stokes Verification for Axial Compressor Flow Analysis. *Proc. 2nd Asian CFD-Conference*, Tokyo, Vol. 2, pp. 151 - 156, 1996

[6] Wiedermann,A.; CFD-Analysis of 3D Viscous Effects in Turbomachinery Cascades Including Tip Clearance Flow. *Proc. Yokohama IGTC*, Yokohama, Vol. 1, pp. 127 - 134, 1995

[7] Kiock, R., Lehthaus, F., Baines, N.C., Sieverding, C.H.; The Transonic Flow Through a Plane Turbine Cascade as Measured in Four European Wind Tunnels. *ASME* 85-GT-44, 1985

[8] Abu-Ghannam, B.J., Shaw, R.; Natural Transition of Boundary Layers - The Effect of Turbulence, Pressure Gradient, and Flow History. *J. of Mechanical Engineering Sciences*, Vol. 22, pp. 213 - 222, 1980

[9] Steinert, W., Eisenberg, B., Starken, H.; Design and Testing of a Controlled Diffusion Airfoil Cascade for Industrial Axial Flow Compressor Application. *ASME* 90-GT-140, 1990

[10] Swanson, R.C., Turkel, E.; On Central Difference and Upwind Schemes. *J. of Computational Physics*, Vol. 101, pp. 292-306, 1992

[11] Strazisar, A.J., Wood, J.R., Hathaway, M.D., Suder, K.L.; Laser Anemometer Measurements in a Transonic Axial-Flow Fan Rotor. *NASA TP* 2879, 1989

[12] Dunker, R.J., Rechter, H., Starken, H., Weyer, H.B.; Redesign and Performance Analysis of a Transonic Axial Compressor Stator and Equivalent Plane Cascade with Subsonic Controlled Diffusion Blades. *ASME* 83-GT-208, 1983

[13] Eguchi, T., Wiedermann, A.; Numerical Analysis of Unstalled and Stalled Flutter Using a Navier-Stokes Code with Deforming Meshes. In *Unsteady Aerodynamics and Aeroelasticity of Turbomachines* (ed.: Y. Tanida, M. Namba), pp. 237 - 254, Elsevier 1995

[14] Dunker, R.J.; Flow Measurements in the Stator Row of a Single-Stage Transonic Axial-Flow Compressor with Controlled Diffusion Stator Blades. *AGARD CP* 351, Paper 23, 1983

[15] Wiedermann, A.; Miyagawa, K., Eguchi, T.; On the Development of Viscous Solvers for Computation of Transient Flows in Turbomachines. To be presented at *ISROMAC-7*, Honolulu, 1998

Flow Simulation in an Aerodynamic Diffusor of a High Loaded Radial Compressor using Different Turbulence Models

M. Heinrich and I. Teipel
Institute for Mechanics, University of Hannover,
Germany

1 Abstract

A two- and three-dimensional code solving the Reynolds-averaged compressible Navier-Stokes equations has been developed and successfully used for computation of the steady flow field in an aerodynamic diffuser of a high loaded centrifugal compressor. For the purpose of comparison the turbulence model of Baldwin and Lomax with and without an extension of Goldberg and Chakravarthy [1] for the determination of separated flow regions and the two equation $k-\epsilon$ model according to Kunz and Lakshminarayana [2, 3] are applied to compute the flow field of the diffuser on several operating points with the two-dimensional code. The three-dimensional solver in addition with the extended and unextended Baldwin-Lomax model has been applied to ascertain the three-dimensional flow field on the working point with the largest separation zone in the two-dimensional case. For comparison measured pressure distributions [4] of the examined operating points are presented likewise.

applied by Compton et al. [6] to an external afterbody flow with jet exhaust. One goal of this paper is to show the properties of this algebraic model by simulating the internal steady transonic flow field of an aerodynamic diffuser with a backflow zone. The computation will be compared with the simulations using the two equation $k-\epsilon$ turbulence model of Kunz and Lakshminarayana for the two-dimensional flow. The $k-\epsilon$ model has been implemented without any wallfunction to reduce empirical influences as much as possible. The range of the Reynolds-number in the investigated cases was 3.064 10^6 up to 4.340 10^6. The reduced mass flux varied between 4.44 kg/s and 9.04 kg/s.
Futhermore the Goldberg-Chakravarthy modification has been prepared for the prediction of separated three-dimensional internal flows. These are the real problems, because flow fields in curved channels are not two-dimensional. The capability of the three-dimensional method will be demonstrated by computing the flow in an aerodynamic diffuser of a high loaded radial compressor. One has chosen for the operation point values which gave for the two-dimensional cases, that have been investigated, the largest separation bubble. The Reynolds-number was 4.340 10^6 and the mass flux 5.49 kg/s. To validate the results of this study distributions of the static pressure have been compared with the experimental data given by Jansen [4].

4 Description of the Numerical Scheme

The developed procedure is an explicit central-finite-difference time-integration scheme. It is first order accurate in time and second order accurate in space. The Reynolds-averaged equations Eq. 4.1 (here 2D version) are formulated in normalized and conservative manner and transformed from the Cartesian coordinate system to a generalized curvilinear system.

$$\frac{\partial\left(\frac{1}{J}\vec{Q}\right)}{\partial t}=\frac{\partial\left[\frac{1}{J}\left(\vec{E}_\nu-\vec{E}_c\right)\right]}{\partial\xi}+\frac{\partial\left[\frac{1}{J}\left(\vec{F}_\nu-\vec{F}_c\right)\right]}{\partial\eta} \tag{4.1}$$

with

$$\vec{Q}=\begin{bmatrix}\bar{\rho}\\ \bar{\rho}\tilde{u}\\ \bar{\rho}\tilde{v}\\ \bar{\rho}\tilde{E}\end{bmatrix};\ \vec{E}_c=\begin{bmatrix}\bar{\rho}\tilde{U}\\ \bar{\rho}\tilde{u}\tilde{U}+\xi_x\bar{p}\\ \bar{\rho}\tilde{v}\tilde{U}+\xi_y\bar{p}\\ (\bar{\rho}\tilde{E}+\bar{p})\tilde{U}\end{bmatrix};\ \vec{F}_c=\begin{bmatrix}\bar{\rho}\tilde{V}\\ \bar{\rho}\tilde{u}\tilde{V}+\eta_x\bar{p}\\ \bar{\rho}\tilde{v}\tilde{V}+\eta_y\bar{p}\\ (\bar{\rho}\tilde{E}+\bar{p})\tilde{V}\end{bmatrix};$$

$$\vec{E}_\nu=\begin{bmatrix}0\\ \xi_x\tau_{xx}+\xi_y\tau_{yx}\\ \xi_x\tau_{xy}+\xi_y\tau_{yy}\\ \xi_x(\tilde{u}\tau_{xx}+\tilde{v}\tau_{xy}-q_x)+\xi_y(\tilde{u}\tau_{yx}+\tilde{v}\tau_{yy}-q_y)\end{bmatrix};$$

$$\vec{F}_\nu = \begin{bmatrix} 0 \\ \eta_x \tau_{xx} + \eta_y \tau_{yx} \\ \eta_x \tau_{xy} + \eta_y \tau_{yy} \\ \eta_x(\tilde{u}\tau_{xx} + \tilde{v}\tau_{xy} - q_x) + \eta_y(\tilde{u}\tau_{yx} + \tilde{v}\tau_{yy} - q_y) \end{bmatrix} ;$$

$$J = \xi_x \eta_y - \eta_x \xi_y ; \quad \tilde{U} = \xi_x \tilde{u} + \xi_y \tilde{v} ; \quad \tilde{V} = \eta_x \tilde{u} + \eta_y \tilde{v} ;$$

$$\tau_{ij} = \frac{\mu_t + \mu_l}{Re} \left[\left(\frac{\partial \tilde{u_i}}{\partial x_j} + \frac{\partial \tilde{u_j}}{\partial x_i} \right) - \frac{2}{3} \delta_{ij} \frac{\partial \tilde{u}_k}{\partial x_k} \right] - \frac{2}{3} \delta_{ij} \bar{\rho} \tilde{k} \quad ;$$

and

$$q_{x_i} = -\frac{\kappa}{\kappa - 1} \frac{1}{Re} \left(\frac{\mu_l}{Pr} + \frac{\mu_t}{Pr_t} \right) \frac{\partial}{\partial x_i} \left(\frac{\overline{p}}{\overline{\rho}} \right) \quad .$$

The discretisized difference equations are solved by a hybrid multistep Runge-Kutta scheme [7]. To prevent the numerical solution from oszillations, especially in the vicinity of shocks and from high-order oscillation modes, a second- and fourth-order artificial dissipation [8] is introduced. To speed up the convergence of the scheme the following acceleration techniques are employed: local time-stepping, implicit smoothing of the residuals with constant or weighted coefficients [9, 10] and a multigrid method [11]. Futhermore the original five-step Runge-Kutta scheme has been reduced to a three-step one according to Leicher [12].

5 Turbulence Models

By solving the Reynolds-averaged Navier-Stokes equations it is necessary to introduce a turbulence model to determine the eddy viscosity, respectively the turbulent viscosity coefficient μ_t in the viscouse flux vectors of Eq. 4.1. Three different models have been applied.

5.1 The Baldwin-Lomax Model

The Baldwin-Lomax model [13] is one of the best known algebraic methods. According to Cebeci [14] the boundary layer is divided up into an inner and an outer layer. Length- and velocity-scale of the eddy viscosity are evaluated directly by means of the flow data and the coordinates of the determined flow point. To predict the turbulent viscosity coefficient $\mu_t = \mu_{t,i}$ in the inner layer Eq. 5.1 is used with $n_b \equiv 0$:

$$\mu_{t,i} = Re\, \overline{\rho}\, \ell^2\, |\omega| \tag{5.1}$$

using

$$|\omega| = \sqrt{\left(\frac{\partial \tilde{v}}{\partial x} - \frac{\partial \tilde{u}}{\partial y}\right)^2} \quad ;$$

$$\ell = k\, n_{BL} \left[1 - exp \frac{-n^+}{A^+}\right] \quad ;$$

$$n^+ = \left(\frac{\sqrt{Re\, \overline{\rho}_w\, \tau_w}}{\mu_w}\right) n_{BL} \quad \text{and} \quad n_{BL} = n - n_b \quad .$$

In the outer layer $\mu_t = \mu_{t,o}$ is calculated by means of Eq. 5.2; n_{BL} and n^+ are defined as in Eq. 5.1. $\vec{w}_{diff}$ is the velocity difference across the boundary layer.

$$\mu_{t,o} = Re\, K\, C_{cp}\, \overline{\rho}\, F_{wake}\, F_{Kleb} \tag{5.2}$$

with

$$F_{wake} = MIN \begin{cases} n_{BL}|_{Fmax}\; F|_{Fmax} \\ C_{wk}\; n_{BL}|_{Fmax}\; \vec{w}^2_{diff} / \; F|_{Fmax} \end{cases} \quad ;$$

$$F = n_{BL} |\omega| \left[1 - exp\left(\frac{-n^+}{A^+}\right)\right] \quad ;$$

$$F_{Kleb} = \left[1 + 5.5 \left(\frac{n_{BL}\, C_{Kleb}}{n_{BL}|_{Fmax}}\right)^6\right]^{-1} \quad ;$$

$$A^+ = 26 \; ; \; C_{cp} = 1.6 \; ; \; C_{Kleb} = 0.3 \; ;$$

$$C_{wk} = 0.25 \; ; \; k = 0.4 \text{ and } K = 0.0168 \; .$$

5.2 Modification of Goldberg and Chakravarthy

Goldberg and Chakravarthy [1] extended the original model by Baldwin and Lomax by adding two new layers to the inner and outer layer in case of a separated flow region: a backflow zone (bubble) and an adaption layer. They suggested a Gaussian distribution of

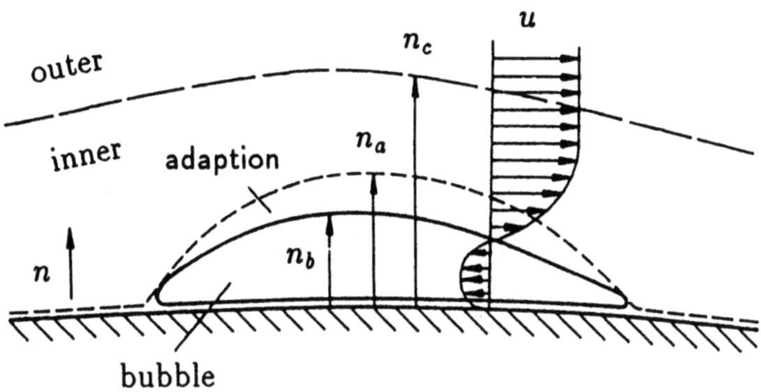

Fig. 5.1: Structure of a Bubble Region along a Wall

the turbulent viscosity in the bubble region and employed the adaption layer to ensure a succesive transition of the eddy viscosity from the back flow regime to the inner layer. The surface pointing to the inner flow in which no velocity parallel to the surface occures (outer bubble surface) is treated as reference wall for the Baldwin-Lomax model application [6]. The viscosity coefficient $\mu_t = \mu_{t,i}$ in the inner layer and $\mu_t = \mu_{t,o}$ in the outer layer is evaluated by means of Eq. 5.1 and 5.2, but now n_b is the related outer bubble surface from the wall (see Fig. **5.1**). In the bubble region Eq. 5.3 is applied:

$$\mu_{t,b} = C_1\, u_s\, n_b \sqrt{\overline{\rho}_w\, \overline{\rho}\, Re}\left(A\left(\frac{n}{n_b}\right) + B\right)\sqrt{G(n)} \tag{5.3}$$

making use of

$$u_s = \sqrt{\frac{\omega|_{\omega_{max}}\, \mu|_{\omega_{max}}}{\rho|_{\omega_{max}}}} \quad ; \quad G(n) = \frac{1 - exp^{-\Phi\left(\frac{n}{n_b}\right)}}{1 - exp^{-\Phi}} \quad ;$$

$$C_{BL} = 0.353 \;\; ; \;\; A = -0.151 \;\; ; \;\; B = 0.684 \text{ and } \Phi = 0.5.$$

To realize an adaption from $\mu_{t,b}$ to $\mu_{t,i}$ the values $\mu_{t,i}$ lower then $\mu_{t,a}$ according to Eq. 5.4 are replaced by $\mu_{t,a}$ as long as the first value of $\mu_{t,i}$ is bigger than $\mu_{t,a}$.

$$\mu_{t,a} = C_1\, u_s\, n_b \sqrt{\overline{\rho}_w\, \overline{\rho}\, Re}\,(A + B)\sqrt{G(n_b)} \tag{5.4}$$

Thus along one wall of a channel the viscouse coefficient μ_t follows Eq. 5.5.

$$\begin{array}{llllll} \mu_t = \mu_{t,b} & \text{while} & 0 & \leq n & < n_b & \text{(bubble)} \\ \mu_t = \mu_{t,a} & \text{while} & n_b & \leq n & < n_a & \text{(adaption layer)} \\ \mu_t = \mu_{t,i} & \text{while} & n_a & \leq n & < n_c & \text{(inner layer)} \\ \mu_t = \mu_{t,o} & \text{while} & n_c & \leq n & & \text{(outer layer)} \end{array} \tag{5.5}$$

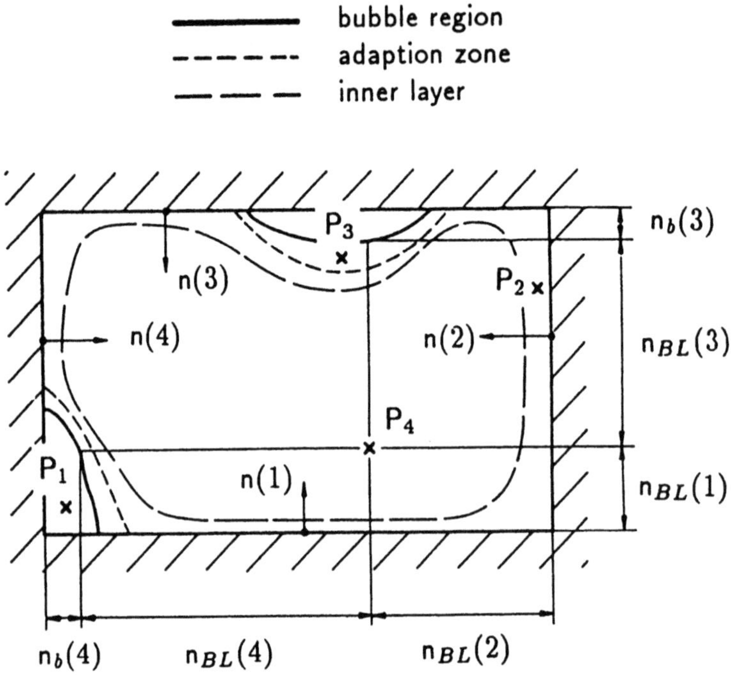

Fig. 5.2: Section perpendicular to the Main Flow Direction of a Channel with Separation Region

5.3 3D Approach of the Algebraic Models

For the computation of multi-dimensional flow structures it is necessary to combine the μ_t-values related to the bounding walls in a suitable manner. In this paper the resulting coefficient to each point in the flow field is determined by weighting the wall related μ_t-values with the inverse related wall distances and summing up all values of the "lowest" related layer type. For the four points in Fig. **5.2** this leads to Eq. 5.6 - 5.9.

Point P_1:

$$\mu_t = \frac{\dfrac{\mu_{t,b}(1)}{n(1)} + \dfrac{\mu_{t,b}(4)}{n(4)}}{\left(\dfrac{1}{n(1)} + \dfrac{1}{n(4)}\right)} \tag{5.6}$$

Point P_2:

$$\mu_t = \frac{\sum_{i=1}^{3} \frac{\mu_{t,i}(i)}{n(i)}}{\sum_{i=1}^{3} \frac{1}{n(i)}} \tag{5.7}$$

Point P_3:

$$\mu_t = \mu_{t,a}(3) \tag{5.8}$$

Point P_4:

$$\mu_t = \frac{\sum_{i=1}^{4} \frac{\mu_{t,0}(i)}{n_{BL}(i)}}{\sum_{i=1}^{4} \frac{1}{n_{BL}(i)}} \tag{5.9}$$

The vorticity-scale used for the prediction of μ_t in the 3D approach is given by the following expression.

$$|\omega| = \sqrt{\left(\frac{\partial\tilde{w}}{\partial y} - \frac{\partial\tilde{v}}{\partial z}\right)^2 + \left(\frac{\partial\tilde{u}}{\partial z} - \frac{\partial\tilde{w}}{\partial x}\right)^2 + \left(\frac{\partial\tilde{v}}{\partial x} - \frac{\partial\tilde{u}}{\partial y}\right)^2} \tag{5.10}$$

5.4 $k - \epsilon$ Model

The employed model was developed by Kunz and Lakshminarayana [2]. It takes care of compressibility effects in a restricted scope only. Density fluctuations and pressure diffusion terms are neglected in the transport equation of the kinetic turbulent energy k and the equation for the turbulent dissipation rate ϵ. This model has been applied successfully in the computation of turbomachinery flows [3]. To ensure a good vectorization level the turbulence transport equations Eq. 5.11 are solved in combination with the other flow equations Eq. 4.1.

$$\frac{\partial\vec{Q}}{\partial t} = \frac{\partial\left(\vec{E}_\nu - \vec{E}_c\right)}{\partial\xi} + \frac{\partial\left(\vec{F}_\nu - \vec{F}_c\right)}{\partial\eta} + \vec{S} \tag{5.11}$$

by means of

$$\vec{Q} = \frac{1}{J}\begin{bmatrix} \bar{\rho}\tilde{k} \\ \bar{\rho}\tilde{\epsilon} \end{bmatrix} \; ; \; \vec{E}_c = \frac{1}{J}\begin{bmatrix} \bar{\rho}\tilde{k}\tilde{U} \\ \bar{\rho}\tilde{\epsilon}\tilde{U} \end{bmatrix} \; ; \; \vec{F}_c = \frac{1}{J}\begin{bmatrix} \bar{\rho}\tilde{k}\tilde{V} \\ \bar{\rho}\tilde{\epsilon}\tilde{V} \end{bmatrix} \; ;$$

$$\vec{E}_\nu = \frac{1}{Re}\frac{1}{J}\begin{bmatrix} \mu_k & \left\{(\nabla\xi\,\nabla\xi)\frac{\partial\tilde{k}}{\partial\xi} + (\nabla\xi\,\nabla\eta)\frac{\partial\tilde{k}}{\partial\eta}\right\} \\ \mu_\epsilon & \left\{(\nabla\xi\,\nabla\xi)\frac{\partial\tilde{\epsilon}}{\partial\xi} + (\nabla\xi\,\nabla\eta)\frac{\partial\tilde{\epsilon}}{\partial\eta}\right\} \end{bmatrix} \; ;$$

$$\vec{F}_\nu = \frac{1}{Re}\frac{1}{J}\begin{bmatrix} \mu_k & \left\{(\nabla\eta\,\nabla\xi)\frac{\partial\tilde{k}}{\partial\xi} + (\nabla\eta\,\nabla\eta)\frac{\partial\tilde{k}}{\partial\eta}\right\} \\ \mu_\epsilon & \left\{(\nabla\eta\,\nabla\xi)\frac{\partial\tilde{\epsilon}}{\partial\xi} + (\nabla\eta\,\nabla\eta)\frac{\partial\tilde{\epsilon}}{\partial\eta}\right\} \end{bmatrix} \; ;$$

$$\mu_k = \mu_l + \frac{\mu_t}{Pr_k} \; ; \; \mu_\epsilon = \mu_l + \frac{\mu_t}{Pr_\epsilon} \; ;$$

$$\vec{S} = \frac{1}{Re}\frac{1}{J}\begin{bmatrix} P - Re\,\overline{\rho}\,\tilde{\epsilon} + \mathcal{D} \\ (C_1\,P - Re\,C_2\,f_2\,\overline{\rho}\,\tilde{\epsilon})\,\frac{\tilde{\epsilon}}{\tilde{k}} + \mathcal{E} \end{bmatrix} \quad ;$$

$$\begin{aligned} P \;=\; & \left\{2\,\mu_t\,\frac{\partial\tilde{u}}{\partial x} - \frac{2}{3}\left[Re\,\overline{\rho}\tilde{k} + \mu_t\left(\frac{\partial\tilde{u}}{\partial x} + \frac{\partial\tilde{v}}{\partial y}\right)\right]\right\}\frac{\partial\tilde{u}}{\partial x} + \\ & \left\{2\,\mu_t\,\frac{\partial\tilde{v}}{\partial y} - \frac{2}{3}\left[Re\,\overline{\rho}\tilde{k} + \mu_t\left(\frac{\partial\tilde{u}}{\partial x} + \frac{\partial\tilde{v}}{\partial y}\right)\right]\right\}\frac{\partial\tilde{v}}{\partial y} + \\ & \mu_t\left(\frac{\partial\tilde{u}}{\partial y} + \frac{\partial\tilde{v}}{\partial x}\right)^2 \quad ; \end{aligned}$$

$$\mathcal{D} = -\frac{2\,\mu_l\,\tilde{k}}{n^2} \quad ; \quad \mathcal{E} = -\frac{2\,\mu_l\,\tilde{\epsilon}}{n^2}\,exp\left(-0.5n^+\right) \quad ;$$

$$f_2 = 1 - \frac{2}{9}\,exp\left(\frac{-R_T^2}{36}\right) \quad ; \quad R_T = Re\,\frac{\overline{\rho}\,\tilde{k}^2}{\mu_l\,\tilde{\epsilon}} \quad ;$$

$$f_\mu = 1 - exp\left(-0.0115\,n^+\right) \quad ; \quad n^+ = \frac{n}{\mu_l}\,\sqrt{Re\,\overline{\rho}\,\tau_w} \quad ;$$

$$Pr_k = 1.0 \; ; \; Pr_\epsilon = 1.3 \; ; \; C_1 = 1.35 \text{ and } C_2 = 1.80 \quad .$$

The turbulent viscosity at last results from a nonlinear relation Eq. 5.12 of k and ϵ with the coefficient $C_\mu = 0.09$.

$$\mu_t = Re\,\frac{C_\mu\,f_\mu\,\overline{\rho}\,\tilde{k}^2}{\tilde{\epsilon}} \tag{5.12}$$

Since it is difficult to "cold start" this model [2], the difference equations for k and ϵ are solved at first in a so called bypass mode by using e.g. the Baldwin-Lomax model to predict a stable flow regime.

5.5 Boundary Conditions

The total values of pressure and density in the inlet section of the diffuser are given according to measurements by Jansen [4]. The tangential velocity is adjusted in a way, that the computed mean-velocity between the leading edges of the diffuser blades fits with measured data. For the 2D simulations single values are given, in the 3D application data profiles perpendicular to the diffuser's frontwall are prescibed.

The radial velocity is evaluated by means of an approach of the theory of characteristics [15]. k and ϵ are predicted by Eq. 5.13 and 5.14 with $Tu_\infty \approx 0.01$, length-scale L_∞ approximately 1 % of the channel width and $\mu_{t,\infty} \approx 0.2\mu_{l,\infty}$.

$$\tilde{k}_\infty = \frac{3}{2}\left(Tu_\infty \tilde{U}_\infty\right)^2 \tag{5.13}$$

$$\tilde{\epsilon}_\infty = Re\,\frac{C_\mu f_\mu \overline{\rho}_\infty \tilde{k}^2_\infty}{\mu_{t,\infty}} \tag{5.14}$$

Along solid walls the first derivative of pressure and density, the velocity, k and ϵ are assumed to be zero. In the diffuser exit surface the static pressure is determined in such a way, that the evaluated static mean-pressure between the trailing edges of the blades fits the pressure measured by Jansen. All other values are extrapolated from the inner flow field. In front of and behind the blades periodicity conditions are used.

6 Results

In **Fig. 6.1** velocity vector plots in the diffuser are shown. They have been computed with three different 2D solvers according to a working point at the choked line at partial load. The whole flow fields predicted with the different solvers are very similar, therefore only the one due to the extended Baldwin-Lomax model has been plotted. Region A shows vectors in vicinity of the flow separation at the pressure surface of the blading. For the Baldwin-Lomax model only a very small separation zone is obtained, whereas the other models yield to larger separation regions. It is remarkable, that the extended Baldwin-Lomax model needs only 49 grid points across the channel to achieve nearly the same resolution as the $k-\epsilon$ model using 97 points. The effect of the larger bubble can be seen in **Fig. 6.2**. Here the static pressure along the meridinal section versus the aspect ratio r/r_o is shown. The shock at $r/r_o \approx 1.3$ appears nearly at the same position, when using the extended Baldwin-Lomax model or the $k-\epsilon$ model. Besides small differences a good agreement of the computed data with the measurement can be seen. The smaller pressure increase downstream of the shock related to the $k-\epsilon$ model in comparison to the increase obtained by the extended Baldwin-Lomax model may be a result of a sightly larger computed separation region in case of the two-equation model.

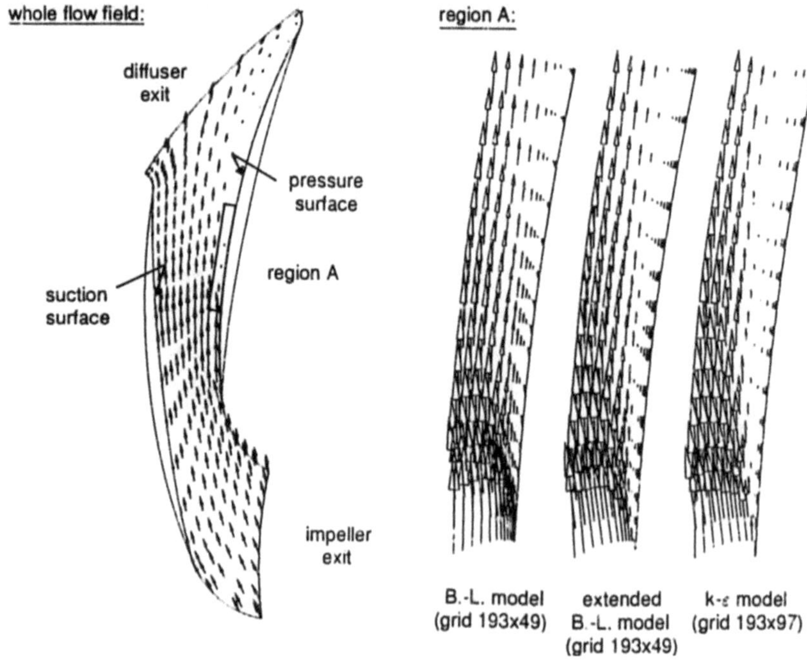

Fig. 6.1: Computed Velocity Vectors in One Diffuser Channel (2D). Working Point at Choked Line (Partial Load): Red. Mass Flux 5.49 kg/s, Red. Impeller Revolution 14 600 1/min.

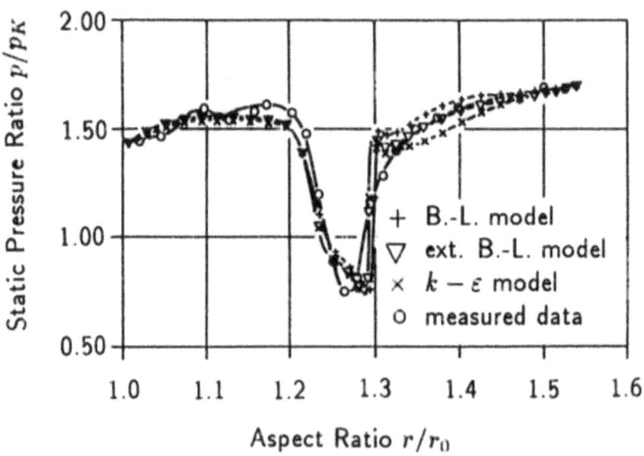

Fig. 6.2: Pressure Ratio p/p_K along Diffuser Meridional Section. Working Point at Choked Line (Partial Load): Red. Mass Flux 5.49 kg/s, Red. Impeller Revolution 14 600 1/min.

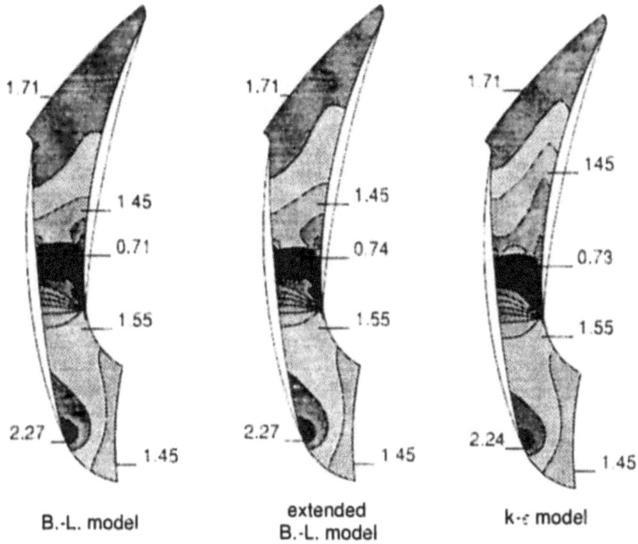

Fig. 6.3: Pressure Ratio p/p_K in a Diffuser Channel (2D Computation). Working Point at Choked Line (Partial Load): Red. Mass Flux 5.49 kg/s, Red. Impeller Revolution 14 600 1/min.

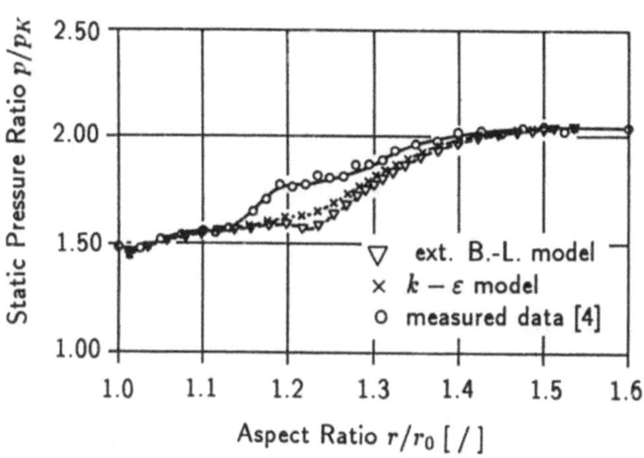

Fig. 6.4: Pressure Ratio p/p_K along Diffuser Meridional Section. Working Point at Surge Line (Partial Load): Red. Mass Flux 4.44 kg/s, Red. Impeller Revolution 14 600 1/min.

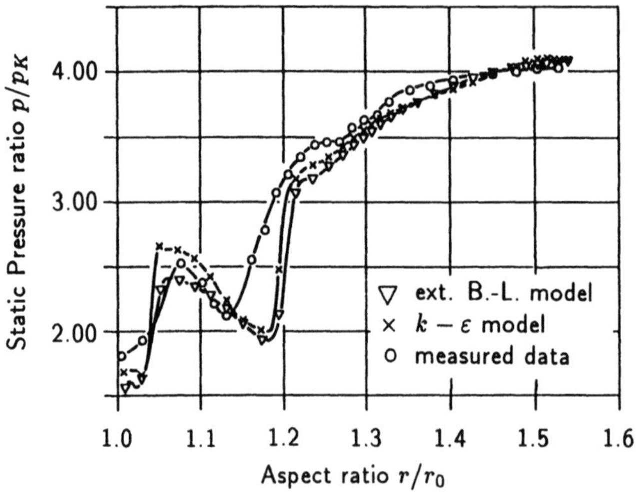

Fig. 6.5: Pressure Ratio p/p_K along Diffuser Meridional Section. Working Point at Surge Line (Full Load): Red. Mass Flux 9.04 kg/s, Red. Impeller Revolution 22 000 1/min.

Distributions of the static pressure ratio p/p_K across the whole width of a diffuser channel obtained by the different turbulence models are presented in **Fig. 6.3**. By comparing **Fig. 6.1** and **6.3**, it is obvious, that the back flow region begins immediately behind the shock. The shock may be identified by means of the "fat" black line downstream the lowest pressure value $p/p_K \approx 0.7$. Furthermore a stronger influence of the bubble towards the pressure fields isobars in the vicinity of the separation zone can be realized when looking at the $k-\epsilon$ model. The distributions according to the Baldwin-Lomax and the extended Baldwin-Lomax model are nearly identical besides the fact of the slightly different shock position.

In **Fig. 6.4** the static pressure ratio distributions along the meridional section are given. They are related to a working point at the surge line at partial load. In this figure are presented only the simulated data related to the extended algebraic model and the two-equation model, besides measured data according to Jansen [4]. Because no separation was detected the computation with the two algebraic models led to an identical solution. Evident the shown simulated data are very similar, too. Therefore they also have in common a too low pressure - compared with the measured data - in the aspect ratio range 1.15 to 1.3.

The diffuser blades start with an aspect ratio r/r_o of 1.15. Considering this value the too low pressure ratios may be a consequence of a too low predicted viscosity in front of

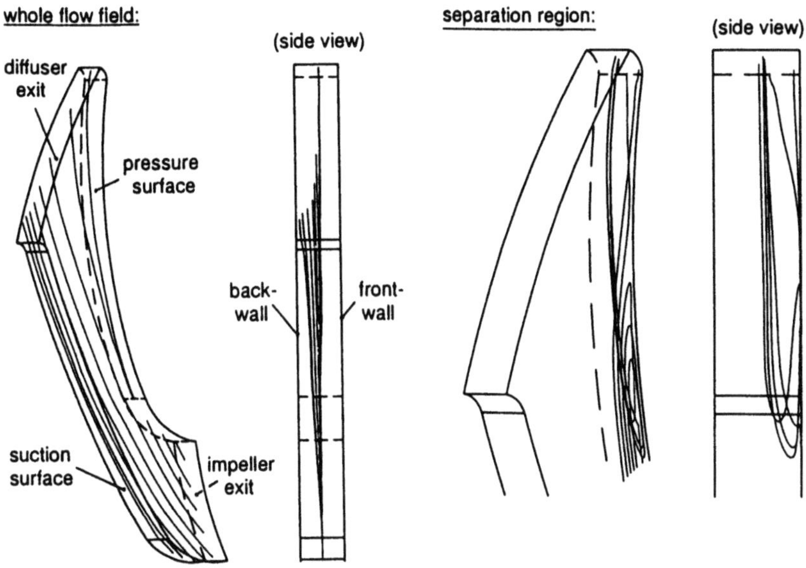

Fig. 6.6: Streamline Distribution in One Diffuser Channel (3D Computation). Working Point at Choked Line (Partial Load): Red. Mass Flux 5.49 kg/s, Red. Impeller Revolution 14 600 1/min.

the leading edge of the blades. This statement may be justified by the curves drawn in **Fig. 6.5**. The figure illustrates the pressure distribution according to a full load working point at the surge limit.

The calculated values differ from the measurement in this figure primary in the section in front of the blades. The fact, that the pressure graphes related to the $k - \epsilon$ model are slightly closer to the measured data, may verify the assertion (of the too low eddy viscosity), too, because the two-equation model predicts the turbulent viscosity coefficient in front of the buckets of the diffuser. The computation is done by means of the distance from leading edge and the needed other data at the blade leading edge. When using the algebraic models the turbulent viscosity coefficient is assumed to be zero in front of the blades. Furthermore the presented results till now are only two-dimensional results, which means, no back- and frontwall influences are taken into consideration.

The quality of the three-dimensional solver using the extended algebraic model is illustrated in **Fig. 6.6** and **6.7**. These figures show three-dimensional results of the working point at the choked line at partial load. The computational grid uses one with 97x37x33 points. In **Fig. 6.6** streamline plots are shown. The streamlines in the upper two graphs show a strong three-dimensional structure of the whole flow field. They all start at the

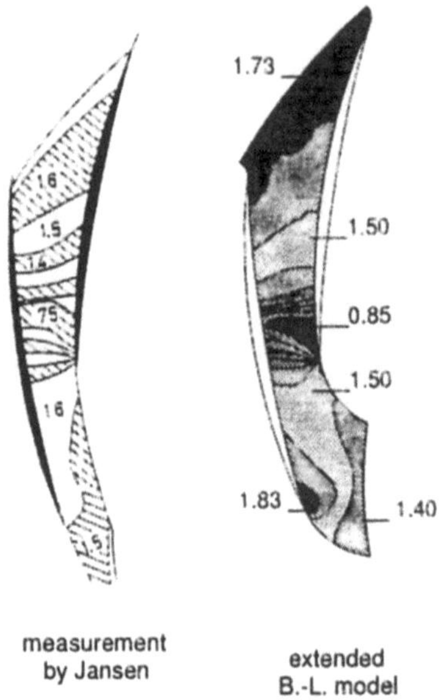

Fig. 6.7: Pressure Ratio p/p_K along the Frontwall of a Diffuser Channel. Working Point at Choked Line (Partial Load): Red. Mass Flux 5.49 kg/s, Red. Impeller Revolution 14 600 1/min.

impeller exit in the center between the diffuser's front- and backwall. But they end in the diffuser exit section between the channel center and the backwall of the diffuser. The two other graphs of the separation region show streamlines which start nearly at the frontwall and end in the exit section at the channel center. The 3D solver in common with the unextended Baldwin-Lomax model yielded to no significant backflow zone.

The static pressure along the frontwall of the diffuser is given in **Fig. 6.7** - in comparison with measured data related to Jansen. In this figure a strong influence of the boundary layer towards the pressure gradient downstream of the pressure minimum can be realized. Futhermore the measured pressure minimum is approximately 13 % smaller. The pressure gradient of the shock in the channel center differs comparatively just slightly from the one detected with the two-dimensional solver. The difference between the simulated reduced mass flux and the measured one was smaller than 3. % and the deviation of the inner efficiency, too.

At last a few words in relation to the convergency criterion, all simulations were driven to a relative deviation of the density (in two succeeding timesteps) about 10^{-6}. The computational time (CPU-time) needed by the two-dimensional approaches were 10 to

50 minutes according to the working point and the used turbulence model, while the three-dimensional simulations required approximately 15 hours.

7 Conclusion

Several working points of an aerodynamic diffuser of a high loaded radial compressor have been simulated using different turbulence models. Simulations using the $k-\epsilon$ model developed by Kunz and Lakshminarayana and computations with the by Goldberg and Chakravarthy modified algebraic Baldwin-Lomax model have shown a good agreement in case of a two-dimensional flow study. When separation regions occur in a flow field, pressure distributions computed with the extended algebraic model are closer to measured data than the data computed with the original model by Baldwin and Lomax. The computation with the modified algebraic model consumed about 4.99 10^{-6} CPU-seconds per iteration and gridpoint for the two-dimensional application on a Fujitsu S400/40 vector-computer with 5 GigaFLOPS peakperformance. The $k-\epsilon$ version required on the other hand about 8.45 10^{-6} CPU-seconds per iteration and gridpoint. Comparing these two values and the results presented above, the extended algebraic model may be an alternative model to the second order closure $k-\epsilon$ model.

The three-dimensional simulation produces a considerable influence of the boundary layer towards the pressure gradient in the boundary layer, which cannot be found in the experiment. Nevertheless the three-dimensional solver is able to predict the flow field in agreement with the measurement. Experimental data are necessary to predict a flow field in coincidence with the reality. This means, e.g., without the correct inlet and outlet conditions the most multi-dimensional simulations are merely academically. Therefore the design of fluid flow engines can only efficient be managed if measurement and computation are working hand in hand. The experiment must give the base data and the simulation than may be used to investigate and understand influences towards the flow field according to parameter and design studies.

References

[1] Goldberg, U. C.; Chakravarthy, S. R.: *Separated Flow Prediction Using a Hybrid k-L / Backflow Model.* AIAA Journal, Vol. 28, NO. 6, Jun. 1990.

[2] Kunz, R.; Lakshminarayana, B.: *Explicit Navier-Stokes Computation of Cascade Flows Using the $k-\varepsilon$ Turbulence Model.* AIAA Journal, Vol. 30, NO. 1, Jan. 1992.

[3] Kunz, R. ; Lakshminarayana, B.: *Three-Dimensional Navier-Stokes Computation of Turbomachinery Flows Using an Explicit Numerical Procedure and a Coupled $k-\varepsilon$ Turbulence Model.* ASME Paper 91-GT-146, 1991.

[4] Jansen, M.: *Untersuchungen an beschaufelten Diffusoren eines hochbelasteten Radialverdichters.* Dissertation, Uni. Hannover, 1982.

[5] Michelassi, V.; Liou, M.-S.; Povinelli, L.A.: *Implicit Solution of Three-Dimensional Internal Turbulent Flows.* NASA TM-103099, 1990.

[6] Compton, W. B.; Abdol-Hamid, K. S. ,Abeyounis, W. K.: *Comparision of Algebraic Turbulence Models for Afterbody Flows with Jet Exhaust.* AIAA Journal, Vol. 30, NO. 1, Nov. 1992.

[7] Radespiel, R.: *A Cell-Vertex Multigrid Method for the Navier-Stokes Equations.* NASA TM-101557, 1989.

[8] Jorgenson, P.; Turkel, E.: *Central Difference TVD and TVB Schemes for Time Dependent and Steady State Problems.* AIAA Paper 92-0053, 1992.

[9] Jameson, A., Baker, T. J.: *Solution of the Euler Equations for Complex Configurations.* AIAA Paper 83-1929, 1983.

[10] Arnone, A.; Liou, M. S.; Povinelli, L. A.: *Multigrid Calculations of Three-Dimensional Viscous Cascade Flows.* ICOMP-91-18, NASA TM 105257; 1991.

[11] Jameson, A.: *MAE Report 1651, 1984.* Published in: Lecture Notes in Mathematics, Bd. 1127, S.156-242, Editor: F. Brezzi, Springer Verlag, 1985.

[12] Leicher, S.: *Numerical Solution of Internal and External Inviscid an Viscous 3-D Flow Fields.* AGARD-CP-412, Vol. 1, 1986.

[13] Baldwin, B. S., Lomax, H.: *Thin Layer Approximation and Algebraic Model for Separated Turbulent Flow.* AIAA Paper 78-257, Jan. 1978.

[14] Cebeci, T.,: *Calculation of Compressible Turbulent Boundary Layers with Heat and Mass Transfer.* AIAA Paper 70-741, 1970.

[15] Chima, R.V.: *Explicit Multigrid Algorithm for Quasi-Three-Dimensional Viscous Flows in Turbomachinery.* Journal Propulsion, Bd. 3, Nr. 5, S.397-405, 1987.

Two Effects of Diaphragm Bulge on Shock Tubes

H. G. Hornung and J. J. Quirk

Graduate Aeronautical Laboratories, California Institute of Technology, Pasadena, California 91125, USA

November 14, 1997

Summary. A computational study is presented that has uncovered two important effects in shock tube flows. They occur because the diaphragm has to support a static load before it breaks. This static load usually deforms the diaphragm before rupture. These effects can have significant consequence both in reflected-shock tunnels and in expansion tubes. The first effect is that the diaphragm bulge causes a shock wave to be propagated across the shock tube and to be repeatedly reflected at the axis and the wall. The second is that the contact surface rolls up to form a vortex ring. The sign of the vorticity in the vortex ring depends on the temperature ratio across the diaphragm.

Key words: shock tube, expansion tube, diaphragm, contact surface

1. Introduction

Shock tubes are devices in which the state of a gas is changed suddenly from one uniform state to another by the passage of a shock wave. For the downstream state to be uniform – a desirable condition in most applications – the shock must be plane and no waves must occur behind it. Usually one argues that, since all waves behind a shock catch up with it, the shock eventually becomes plane. However, in practical situations one does not want to make the diaphragm at which the shock is generated unnecessarily strong, in order to minimize the diaphragm opening time. Therefore the diaphragm, which supports a static pressure difference before rupture, is invariably bulged to some extent.

This problem is studied here by computing the effect of the diaphragm bulge on the ensuing flow. In this study the very simple approach is taken of considering the flow to be inviscid. The diaphragm is assumed to open in zero time. These simplifications are made in the expectation that the effects of diaphragm bulge are qualitatively significant, so that, even in these idealized conditions, important features might well show up.

2. The computational setup

The axisymmetric Euler equations are integrated numerically using a flux-limited, operator-split, Roe solver, as implemented in the *Amrita* computational system, see Appendix A. The shock tube is discretized using a uniform grid of 40 by 360 mesh cells, to which two tiers of local refinement are added to increase the effective computational resolution to 360 cells along the radius and 3240 cells along the axis.

Fig. 1. Pseudo-schlieren image of the flow just after the diaphragm has broken. The shock wave, the contact surface and the expansion wave all still have the approximate shape of the initially imposed pressure discontinuity. Tiny shock waves have begun to form at the walls. Temperature ratio = 1. The imaging technique makes the greyshading a prescribed monotonic function of the magnitude of the density gradient divided by the local density.

The diaphragm is assumed to be supported around its circumference and deformed into a half sine wave, with maximum deflection at the axis equal to 6.25 % of the diameter. The initial conditions chosen are that a pressure discontinuity exists at the line representing the deformed diaphragm, with pressure ratio 20, and that the two regions are separately uniform. The flow is then allowed to evolve, sending a shock into the low-pressure test gas to the right and an expansion into the high-pressure driver gas to the left, with a contact surface separating the two. Fig. 1 shows a pseudo-schlieren image of the flow just after the diaphragm has "broken", sending a shock to the right, and an expansion wave to the left. At the condition shown in Fig. 1, the waves still have approximately the same shape as the original diaphragm. Both gases are assumed to be diatomic with specific heat ratio $\gamma = 1.4$.

3. Early phase of the flow

At the wall, the pressure discontinuity makes a finite angle with the boundary, while at the axis it is normal to the axis. When the diaphragm breaks, a shock wave is formed that initially has the same shape as the diaphragm, and propagates to the right. At the same time, an expansion wave is formed on the left, also with the same initial shape as the diaphragm. The expansion wave propagates to the left and disperses. The situation after a short time is shown in the schlieren image of Fig. 2.

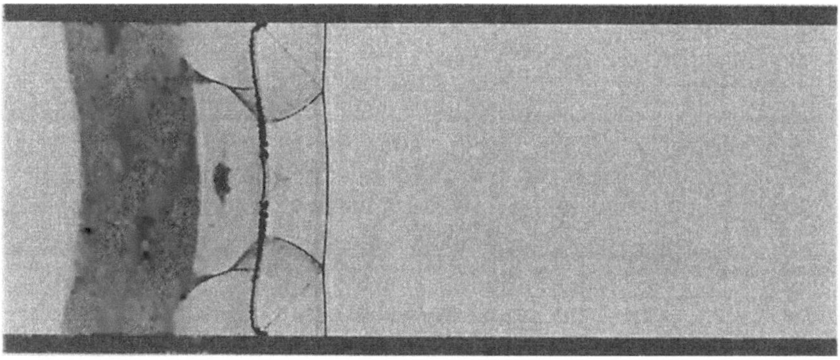

Fig. 2. Same flow as in Fig. 1, a little later. The shock waves that were formed at the walls have propagated a small distance toward the axis.

The shock waves that had just formed near the walls in Fig. 1 can be seen clearly in Fig. 2. They are formed because of the angle between the pressure discontinuity and the wall. The fact that the shock is not normal to the wall causes the flow through it to be deflected toward the wall. This in turn causes a sudden increase in pressure in this region, with the consequence that a shock wave propagates away from it toward the axis.

A similar feature appears in the wall region on the other side of the interface between the driver gas and the test gas. The expansion wave, which is initially thin, is also oblique to the wall. It will therefore also deflect the flow through it toward the wall. The consequence is again the same, a shock is caused to propagate toward the axis.

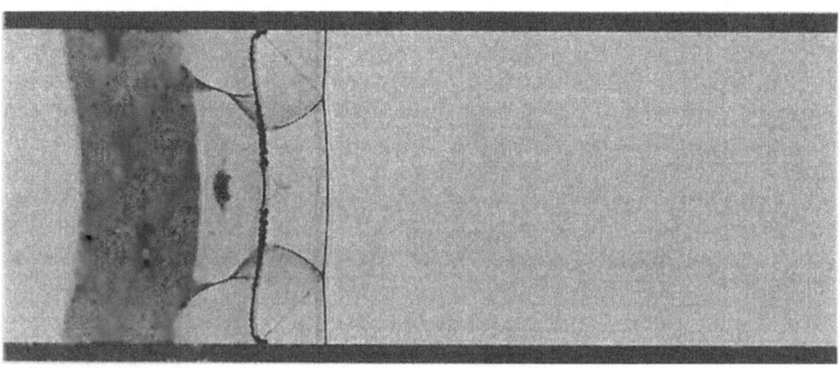

Fig. 3. Further evolution of the flow to the point where the shocks have progressed through approximately half a tube radius. The triple point formed by the shock in the test section with the primary shock generates a shear layer. The contact surface begins to roll forward near the walls.

The shock wave that travels radially inward in the test gas forms a triple point with the primary shock wave. A shear layer issues from this triple point. Both of these features may be seen in Fig. 3. In the driver gas, the shock has progressed through approximately half a tube radius at this time, and the expansion wave has dispersed to a thickness of approximately one tube radius.

4. Further evolution of the flow

Fig. 4 shows the flow a little later, when the shock in the test gas has been reflected from the axis and has travelled approximately half way out to the wall. Note how this shock penetrates the interface between test gas and driver gas, and appears in the driver gas as a precursor. The triple point formed by the driver-gas shock and the primary shock continues to produce a shear layer. On the other side of the interface, the shock in the driver gas is just being reflected at the axis at this time.

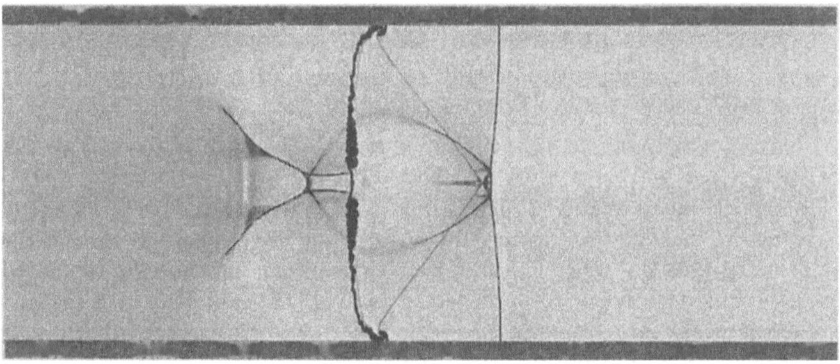

Fig. 4. Further evolution of the flow. The shock in the test gas has been reflected at the axis and is moving back out to the walls. The triple point keeps painting vortex sheets into the test gas. The shock in the driver gas is just being reflected at the axis. Further roll-up of the contact surface has occurred.

The interface has started to develop some instabilities that roll up into eddies. Since no intrinsic length scale exists in the Euler equations, the scale of these eddies must be regarded as a feature which is grid-resolution dependent, but the gross structure of the interface is likely correct. At the wall, the interface is moving forward faster than at the axis. A little later, the shock reflection at the axis in the driver gas has changed to a Mach reflection, see Fig. 5.

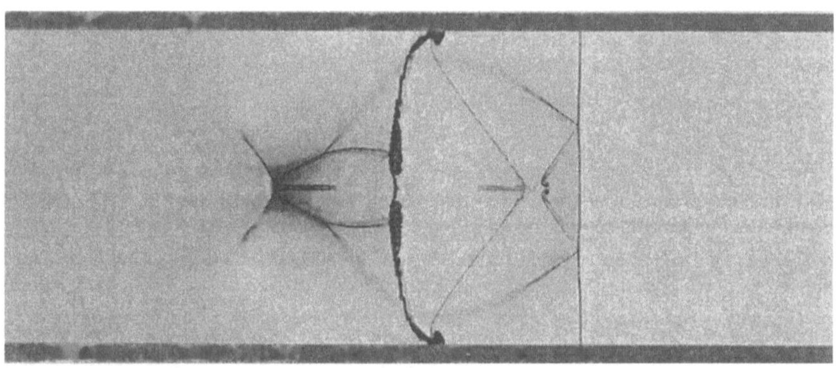

Fig. 5. Further evolution of the flow. The shock in the driver gas has formed a Mach reflection at the axis. The eddies formed on the contact surface are growing.

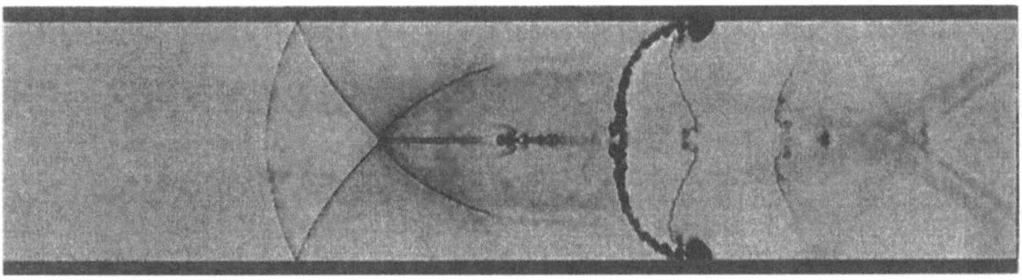

Fig. 6. Much later phase of the flow, when the primary shock has travelled approximately 4 tube diameters and has just reached the right hand end of the tube, which coincides with the edge of the image. Note the strong forward wall jets that have been formed by the contact surface. Note also the waves, shear layers and small vortex rings that are still present in both the test gas and the driver gas.

A much later stage of the flow is shown in Fig. 6, in which the primary shock has just reached the shock tube end, which corresponds to the right-hand edge of the image. Clearly, by this stage, the contact surface has become even more strongly concave, and numerous non-uniformities still exist in both the test gas and the driver gas.

To illustrate the strength of the features described, Fig. 7 shows a composite figure with plots of density profiles in two stages of the flow. The images show the upper half of the tube and two horizontal lines, one near the axis and one near the wall. The density profiles along these two lines are plotted in the two figures underneath each image. The two profiles highlight the differences between the near-wall and near-axis regions and show the size of the density jumps across the waves and contact surfaces.

5. Effect of temperature ratio

Up to this point, all computations shown were for the case of a pressure ratio of 20, and a temperature ratio of 1 across the diaphragm. In many applications, shock tubes are operated with a driver gas speed of sound that is considerably higher than that of the test gas, in order to increase the primary-shock Mach number. This is achieved, either by using a light driver gas, or by raising the driver-gas temperature, or both. In order to examine the influence this would have on the phenomena discussed, another computation was made in which the temperature of the driver gas was 9 times that of the test gas. Fig. 8 shows a sequence of the evolution, and Fig. 9 shows a frame at a later stage. Most of the features of this flow are qualitatively the same as in the case with temperature ratio 1. The striking difference produced by raising the temperature ratio is that the interface is now convex. This is good for reflected-shock tunnels, because it would oppose the wall jet that occurs in viscous flows. Reflected-shock tunnels usually operate with heated driver gas.

6. Implications for practical applications of shock tubes

It is clear from the images and profiles shown in the last section that the flow between the shock and the contact surface is far from uniform. It is criss-crossed

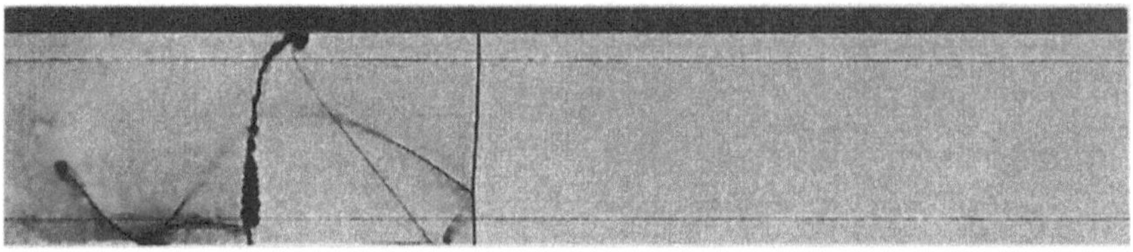

$t = 47.34$

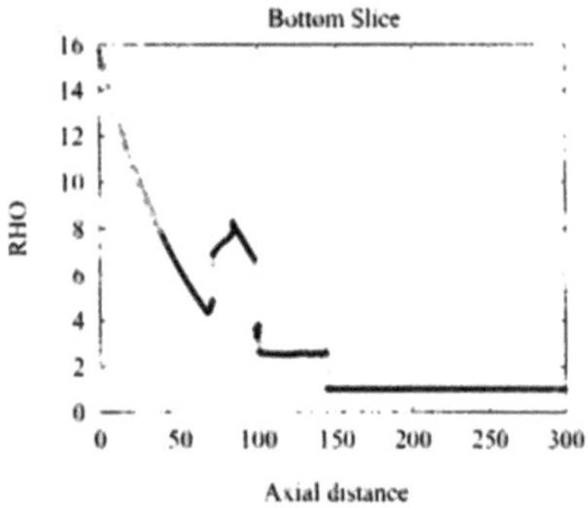

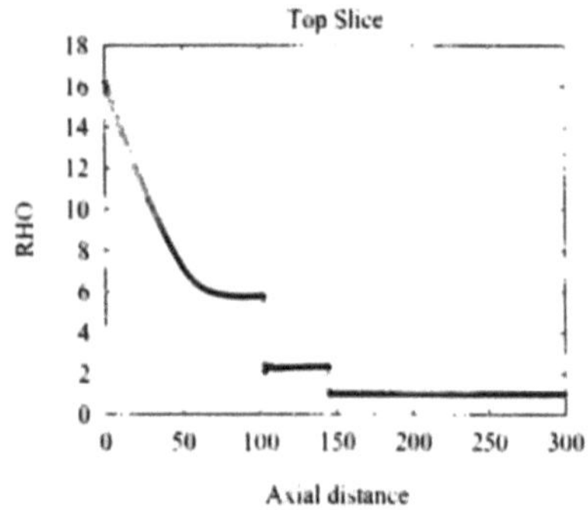

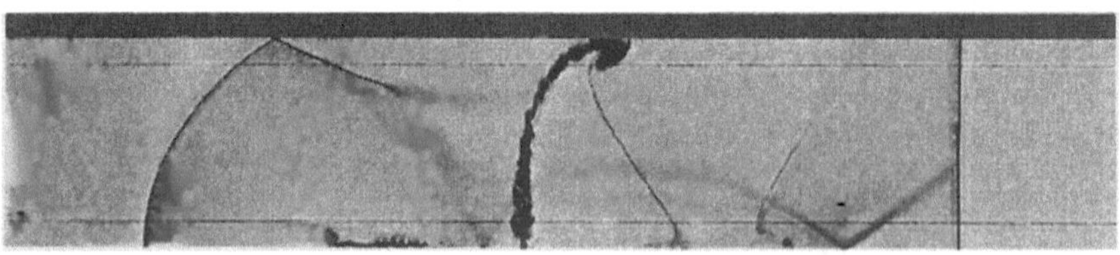

$t = 92.47$

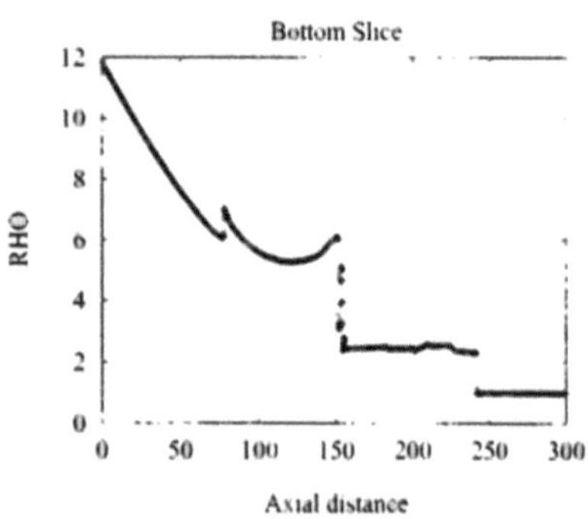

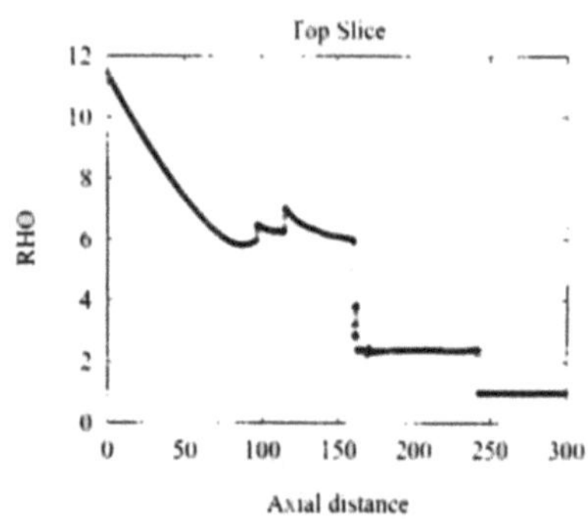

Fig. 7. Composite figure showing two phases of the flow in the form of a pseudo-schlieren image of the upper half of the tube and two density profiles along the lines marked in the images. Note the strength of the density jumps across the features in the flow and the differences between the near-wall and near-axis profiles.

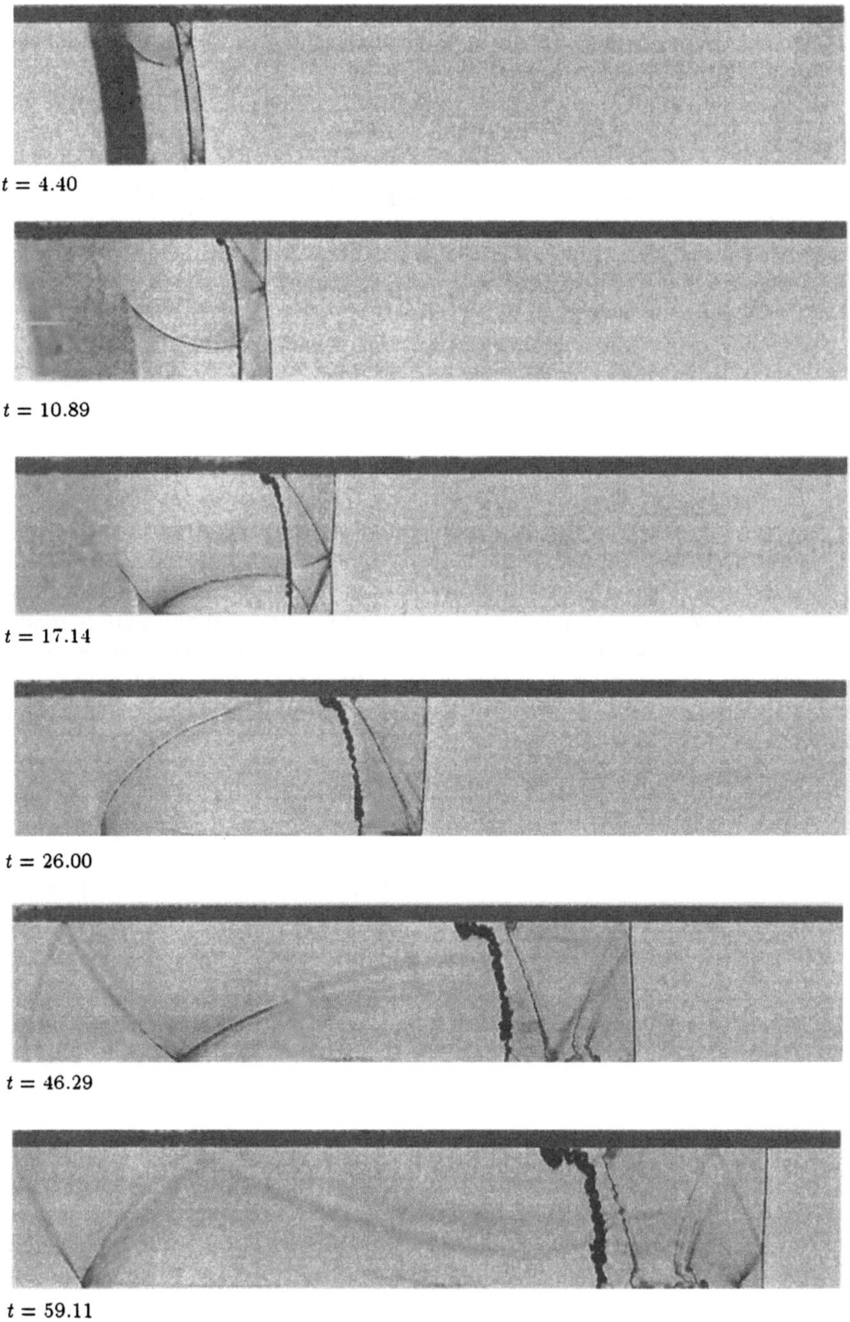

Fig. 8. Sequence showing the evolution of the flow with temperature ratio 9, and pressure ratio 20. Note that the time intervals between the phases shown are not equal. With the high-temperature driver gas, the contact surface becomes concave, in contrast to the case of the cold driver. Also, the transverse shock in the driver is faster than in the test gas, so that the pressure rise across it transmits itself across the interface and causes a precursor shock to be formed there.

with weak shocks and shear layers. This must be kept in mind when the shock tube is desired to produce a uniform flow for the study of chemical reaction rates or gasdynamical investigations.

In another application, the reflected-shock tunnel, the primary shock is reflected from the end of the shock tube to produce what is normally regarded as a uniform reservoir region to be expanded through a convergent-divergent nozzle to produce a uniform free stream. Clearly, such a reservoir would also contain non-uniformities that trace back to the diaphragm bulge. Fortunately, in both cases, these non-uniformities decrease with time, *i. e.* with shock tube length. Shock tubes are typically 50 diameters long, compared with the 4.5 diameters of our computational shock tube.

A second effect that is important for reflected-shock tunnels is the fact that the interface becomes concave, see Fig. 6. This causes the driver-gas to flow forward in a wall jet which is a common cause of early driver-gas contamination in reflected-shock tunnels. It occurs in real flows by the mechanism of shock wave boundary layer interaction, with separation of the boundary layer generated by the incident shock. Here, the concavity of the contact surface resulting from diaphragm bulge aggravates this problem for temperature ratios near 1.

In yet a further application of shock tubes, the flow to the left of the interface is used as the test gas in a two-stage shock tube. This application is the expansion tube. A feature of expansion tubes is that the secondary diaphragm is made as thin and as light as possible, because the performance of the device depends critically on fast opening. Consequently, a diaphragm supported only from its edge will always bulge significantly. Here, the non-uniformities illustrated in the previous section are clearly very important.

Fig. 9. Late phase with temperature ratio 9, showing the convex interface and still deformed shock wave.

7. Conclusions

In a computational study of the flow that ensues after an initially bulged, idealized diaphragm ruptures, two possibly important effects resulting from the bulge are found. One is that transverse shocks reflect alternatively at the axis and the wall, causing the flow in both the test gas and the driver gas to be non-uniform. The non-uniformity decreases with time, so that it may be unimportant in sufficiently long shock tubes. The second is that the interface between test gas and driver gas becomes curved, forming a concave shape when the density of the driver gas is much higher than that of the test gas, and a convex one when the densities are of the same order.

Acknowledgement. Amrita was developed with support by Los Alamos National Laboratory - subcontract 319AP0016-3L under DOE Contract W-7405-ENG-36.

A. *Amrita* – Adaptive Mesh Refinement Interactive Teaching Aid

Amrita is a software system for automating numerical investigations. The base configuration, which is freely available from **http://www.amrita-cfd.com**, consists of a language interpreter called *Amrita* and a plugin computational engine called *Amr_sol*, see Quirk 1997 for details. Note the typographic distinction between *Amrita* the interpreter and *Amrita* the overall system. *Amrita* forms a programmable interface between the user who wishes to simulate a flow problem and the compute engine which has the expertise to carry out the required tasks: generating the grid, applying the boundary conditions, integrating the flow solution, post-processing results etc. By design, the scripting language (*Amrita*) does nothing more than schedule tasks which the computational engine (here *Amr_sol*) follows slavishly. The coupling between the two components is purposefully loose to allow *Amr_sol* to be replaced by other compute engines as the need arises. Thus users can concentrate on composing scientific investigations without worrying unduly about the underlying computational methodology.

The wisdom of endowing *Amrita* with its own programming language might not be immediately obvious, especially to reluctant programmers, but it does allow the system to operate efficiently at several levels of sophistication, thereby allowing an individual to find his or her own level of programming comfort. Moreover, *Amrita* is designed to fill a niche which is not well catered for by existing computer languages. Specifically, *Amrita* is a document preparation language which facilitates the archiving of entire numerical investigations, as distinct from isolated computations. Once archived, an investigation can later be reproduced by any interested party, and not just the original investigator, for no cost other than the raw CPU time needed to process the archived script. For example, this entire paper was produced by a single *Amrita* script. To do this, the script needed to: construct a Roe solver for *Amr_sol* (see below); setup the bulged-diaphragm problem; integrate the axisymmetric Euler equations; produce pseudo-schlieren images of the results; generate LaTeXto typeset this page.

The advantage *Amrita* offers over the traditional archiving of scientific papers and computational datasets is that it allows one researcher to springboard directly off the efforts of another, and frees the follow-up researcher to concentrate solely on those aspects of the original investigation which he or she finds interesting. For instance, someone interested in flow visualization could customize the routine which produced the present pseudo-schlieren images:

```
proc PseudoSchlieren {
     exposure [0:1] = 0.8  # darkness of image
     amp      [0:?] =  15  # magnification of weak features
     grid           = {G}  # subset of computational grid
   }
   DrhoDx      ::= (RHO[+i]-RHO[-i])/(X[+i]-X[-i])
   DrhoDy      ::= (RHO[+j]-RHO[-j])/(Y[+j]-Y[-j])
   RHObar      ::= (RHO[+i]+RHO[-i]+RHO[+j]+RHO[-j])/4.
   schlieren   ::= sqrt(DrhoDx[]**2+DrhoDy[]**2)/RHObar[]
   minmax schlieren[] -> min, max
   wt          ::= (schlieren[]-$min)/($max-$min)
   shading     ::= $exposure*exp(-$amp*wt[])
   plot image $grid m<shading[]>
end proc
```

Side-stepping the syntactic details, the above routine does nothing more than define a symbolic expression for the command **plot image** to render over some selected subset of the computational grid: m<0> is shaded as black and m<1> is shaded as white. The straightforward change:

```
plot image $grid hls<360*shading[],0.5,0.5>
```

would result in images rendered in hue-lightness-saturation colour space. Alternatively, substituting this definition of schlieren[]:

```
DrhoDx      ::= (RHO[+i]-RHO[-i])/(X[+i]-X[-i])
DrhoDy      ::= (RHO[+j]-RHO[-j])/(Y[+j]-Y[-j])
theta       ::= $theta*PI/180.0
schlieren   ::= DrhoDx[]*cos(theta[])+DrhoDy[]*sin(theta[])
```

would result in images which reproduce the directional bias of a real schlieren optical system.

Because *Amrita* is a programmable system, it has no fixed capabilities and should not be mistaken for a CFD code. In fact, this *Amrita* script:

```
EulerEquations
module amr_sol
BasicCodeGenerator {
   solver    = roe_fl
   scheme    = flux-limited'operator-split
   document  = yes
}
```

is sufficient to construct the Roe code used for the present investigation. It generates: the raw *Fortran* source (a lean 351 lines); an *Amrita* script to compile the source; a compiled shared object to link with the module *Amr_sol*; a 30+ page *PostScript* manual which dissects the *Fortran* source, subroutine by subroutine.

As its name suggests, BasicCodeGenerator (BCG for short) has no pretensions to representing the last word in solver sophistication. But as the present results demonstrate, it does produce a serviceable product. Moreover, CFD die-hards can use the output from *BCG* as a template with which to build their own *Amr_sol* compatible code from the ground up. *Amrita* reduces the need for individuals to reinvent the CFD wheel, but it has not been designed as a black-box system which mysteriously produces results upon demand.

This appendix covers but a small part of the current implementation of *Amrita*, and an even smaller part of the underlying philosophy. A more detailed description of the system plus an installation kit is available from http://www.amrita-cfd.com.

References

Gaydon AG and Hurle IR (1963) The shock tube in high-temperature chemical physics. Reinhold Publishing Corporation, New York

Quirk JJ (1997) An Introduction to *Amrita* (in preparation), see also the web site at http://www.amrita-cfd.com

This article was processed by *Amrita* using the LaTeX style file *cljour1* from Springer-Verlag.

Boundary Condition for Pipe Elements in One-Dimensional Numerical Scheme

Masaki ENDO[1] and Junjiro IWAMOTO[2]

[1] Department of Mechanical Engineering, Tokyo Metropolitan College of Technology
1-10-40 Higashi-Ohi, Shinagawa-ku, Tokyo 140, Japan

[2] Department of Mechanical Engineering, Tokyo Denki Unuversity
2-2 Kanda Nishiki-cho, Chiyoda-ku, Tokyo 101, Japan

1 Introduction

The complicated pipe system is widely used in industries with many pipe elements, where the flow is often unsteady and compressible associated with complicated behavior of pressure waves and shock waves. One-dimensional numerical schemes have been applied so far to analyze the unsteady flow field in the pipe. To the best of our knowledge, the numerical simulation was started with Jenny's works(1) where a wave-diagram method was primarily employed. Subsequently, it is proved through many works(2,3) that the numerical analysis of such flow field in the pipe under the assumption of one-dimensional flow is effective and satisfactory in obtaining the unsteady flow pattern in the pipe. Various boundary conditions have been proposed and used to solve numerically the equations governing such a flow field. In general, the results by calculation depend more on boundary condition than on numerical scheme employed. Especially, the boundary conditions play an important role in analyzing the flow field throughout the pipe when a lower level scheme such as a one-dimensional scheme is used, where the empirical coefficients are significant parameters. Consequently, coding the program to analyze the flow field in a complicated pipe system will be very tedious even if a simple scheme is applied. In this paper, the boundary condition for three-way junction is derived based on the wave motion in the junction without the empirical coefficient. And the different boundary conditions for different pipe elements such as a nozzle, an elbow and a bend are generalized to give a unified conditions which are derived by analyzing the wave motion and developing the boundary condition for three-way junction.

2 Numerical Scheme

In order to show the usefulness of the derived boundary condition a Random-Choice Method is applied to the unsteady one-dimensional flow in the pipe and the results are compared to those by experiment. The RCM is well-known to be very useful to analyze the flow field in a pipe under the assumption of one-dimensional flow. The method consists of a sequence of Riemann problems formed in neighboring grids as shown in Fig. 1. The Riemann solver is applied to a certain time level and then the solution at the next time level at the center of the medium between neighboring grids is sought, and so the numerical grid points are alternatively generated as time proceeds. In seeking each solution, pseudo-random number is adopted to keep the physical discontinuities such as the shock wave discontinuous and thus, the locations of the discontinuities fluctuate with time.

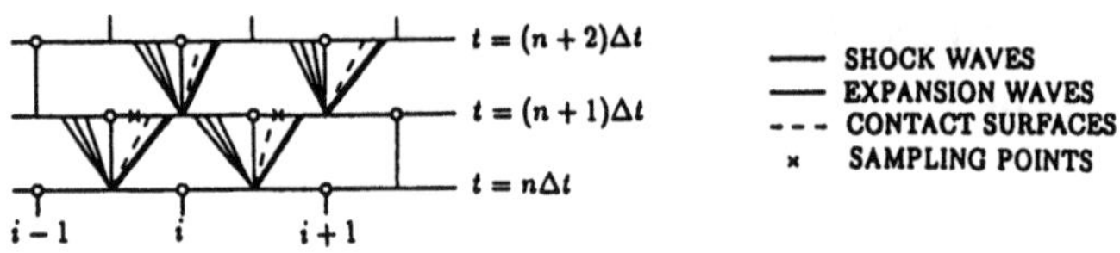

Figure 1. Illustration of Concept for RCM

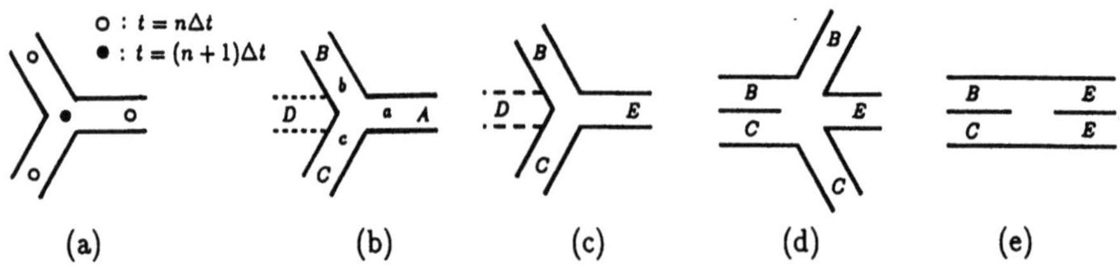

Figure 2. Boundary Conditions for Three-way Junction

2.1 Junction

The boundary condition at a junction is derived based on the motion of the pressure waves through the junction[4]. The geometry of the junction shown in Fig. 2(a) is considered where the angles between two of any pipes are equal to each other i.e. $\frac{2}{3}\pi$ and the cross-sectional area of each pipe is the same. In Fig. 2(a) empty points at the position a half grid length away from the center of junction, i.e. the solid point, indicate the locations where flow states are given at time $t = n\Delta t$ and the flow state at the solid grid point at time $t = (n+1)\Delta t$ is to be calculated using that at three empty grid points. However, it is difficult to apply the RCM directly to the flow field in the junction. And so, the flow state at the solid point is assumed to be represented by three flow states a, b and c as shown in Fig. 2(b) and letters A, B and C refer to the flow states at the empty points, where the flow orientation to the center of junction is assumed to be positive. The flow state at the solid point is assumed to be given by the solution concerning the flow state obtained from the combination of two of any empty points and the flow state at the remaining point.

For instance, the flow state a at the solid point is estimated using the flow state A and the imaginary state D obtained from the combination of B and C. If the two flow states B and C are equivalent to each other, these are combined to give the same flow state as the original one. On the other hand, the flow state at any empty points can be divided into the same flow states at two points. But actually in almost all cases each flow state in two pipes is not equivalent, and so, the general model where the non-equivalent states of the flow through two pipes are combined is given as follows. A sequence of Figs. 2(c) to 2(e) is schematic view which shows how the state of the flow through the junction is reduced to those through the straight pipe and how the flow states B and C are replaced by imaginary flow state D. If the flow state E is assumed to be equal to D, the reflective boundary conditions can be formed between the flow states E and the combined flow

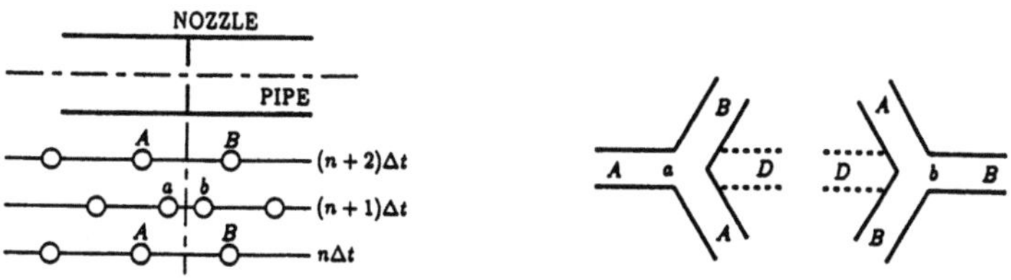

(a) Flow State across Nozzle (b) Imaginary Junction

Figure 3. Boundary Conditions at Nozzle

state obtained from two flow states B and C. First, the flow states B and C in Fig. 2(c) can be divided into two same flow states as shown in Fig. 2(d). Then, the flow states B and C are combined into the flow state E following the definition of E as shown in Fig. 2(c), resulting in the states in the flow through the straight pipe as shown in Fig. 2(e) where the reflective boundary is formed between combined flow state B and C and two Es. Finally, considering the definition of E, the combination of the flow states B and C become equivalent to two Ds. Thus, the flow states through the junction are reduced to those through the single straight pipe so that the imaginary flow state D is evaluated through the following equations of continuity, momentum and energy valid for the flow through the single straight pipe;

$$k_b\rho_b u_b A_b + k_c\rho_c u_c A_c = 2\rho_d u_d A_d \tag{1}$$

$$k_b\rho_b u_b^2 A_b + p_b A_b + k_c\rho_c u_c^2 A_c + p_c A_c = 2\rho_d u_d^2 A_d + 2p_d A_d \tag{2}$$

$$k_b\rho_b u_b A_b\left(\frac{u_b^2}{2}+\frac{c_b^2}{\gamma-1}\right) + k_c\rho_c u_c A_c\left(\frac{u_c^2}{2}+\frac{c_c^2}{\gamma-1}\right) = 2\rho_d u_d A_d\left(\frac{u_d^2}{2}+\frac{c_d^2}{\gamma-1}\right) \tag{3}$$

$$A_b + A_c = 2A_d \tag{4}$$

$$k_b = 1, -1 \qquad k_c = 1, -1 \tag{5}$$

where p, ρ, u, γ and c are the pressure, density, flow velocity, ratio of specific heats and acoustic speed, respectively, and the subscripts b, c and d refer to flow states B, C and D. A is the cross-sectional area of the pipe at the junction and, in this case $A_b = A_c = A_d$. k is positive or negative depending on the flow orientation through each pipe. The flow states at the points b and c are estimated following the same steps as those for the flow state a shown above.

2.2 Nozzle

Schematic illustration about how the flow field near the nozzle is replaced by that near the junction is shown in Fig. 3. As shown in Fig. 3(a), A and B refer to the flow states in the pipe upstream and downstream of the nozzle and a and b the flow states at the position immediately upstream and downstream of the nozzle. These flow states a and b are assumed to represent boundary conditions near the nozzle in the similar way as those for the junction.

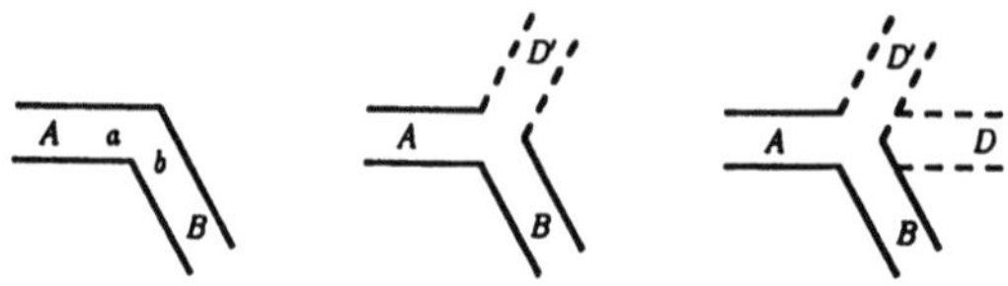

Figure 4. Boundary Conditions for Elbow

Consequently, the flow states a and b at the nozzle should be estimated from the flow states A and B. If the nozzle area ratio to the pipe is equal to 0, i.e. the closed end, the flow state a is estimated independent of the flow state B because of the reflective boundary, while in the flow through the pipe without the nozzle the flow states a and b which are equal to each other are directly evaluated from the flow states A and B. And so, the boundary at the nozzle plate is considered to be reflective and the flow through the opening of the nozzle to be transmitting if the three-dimensional effect of the flow across the nozzle can be ignored. Thus, the flow states a and b across the nozzle may be considered to be obtained by the combination of the flow states at the reflective boundary and those at the free boundary. In estimating the flow state a, the reflective boundary condition at the nozzle is formed by replacing the flow state B in the other side of the nozzle by the flow state A, where the orientation of flow velocity to the nozzle is positive. Therefore, the imaginary junction as shown in Fig. 3(b) is constructed and then, the flow state at time $t = (n+1)\Delta t$ can be obtained. In applying the boundary conditions for the junction to the flow field shown above, Eq.(4) is modified using the area ratio of the nozzle to the pipe ϕ. That is

$$A_d = (1-\phi)A_a + \phi A_b \tag{6}$$

The flow state b is also estimated through the similar procedure as follows.

$$A_d = \phi A_a + (1-\phi)A_b \tag{7}$$

Additionally, it is noted that the boundary condition for the nozzle can be applied to that for the sudden area change by using Eqs.(6) and (7) alternatively.

2.3 Elbow

The boundary conditions for the elbow with the turning angle $\omega = \frac{1}{3}\pi$ can be specified as follows. The flow field through the elbow can be replaced by that through the junction as shown in Fig. 4. Figure 4 shows a sequence of steps for the reduction of the flow field through the elbow into that through a junction, where A and B refer to the flow states in the pipe upstream and downstream of the elbow and a and b the flow states at the position immediately upstream and downstream of the elbow. The flow states through two of any pipes of the junction correspond to those through the elbow. The flow state D' through the imaginary pipe of junction as shown in Fig. 4(b) is estimated by combining

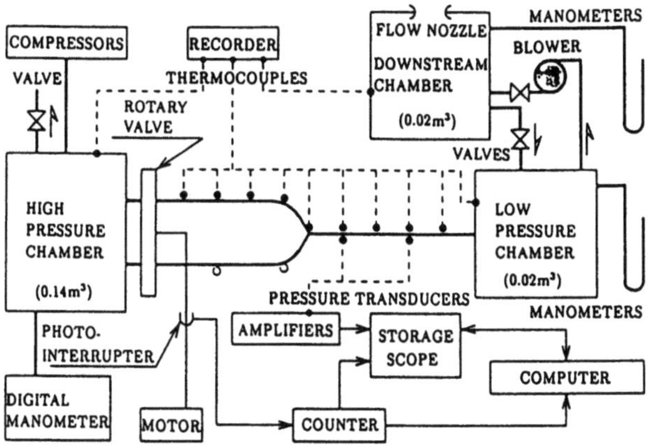

Figure 5. View of Experimental Apparatus

of those through two pipes of the elbow. i.e.

$$A_{d'} = \frac{1}{2}A_a + \frac{1}{2}A_b \tag{8}$$

so that the reflective boundary is constructed between D' and the combination of A and B. Subsequently, the same procedures as those for the junction are taken. And then,

$$A_d = \frac{1}{2}A_{d'} + \frac{1}{2}A_b = \frac{1}{4}A_a + \frac{3}{4}A_b \tag{9}$$

In the estimation of the flow state b the same numerical steps are applied through the similar procedure.

3 Application

In order to show usefulness of the derived boundary conditions for pipe elements such as the junction, the nozzle and the elbow, pulsating flow is experimentally generated in a pipe with each pipe element and the results are compared to those by the RCM.

3.1 Experimental Considerations

Figure 5 shows schematically a view of the experimental apparatus. The compressed air is steadily supplied into a high pressure chamber where the pressure is kept at 4 atmospheres by the control valve. The compressed air in the high pressure chamber is discharged into a pipe through the rotary valve by a convergent nozzle, resulting in pulsating flow through the pipe. The pipe discharges the flow into a low pressure chamber where the pressure is kept atmospheric using a blower. The rotary valve is composed of the rotary disc and the casing as shown in Fig. 6. The casing has either two or one port according to the type of the pipe, i.e. the number of the intlets of the pipe system, and the rotary disc has two openings. The temporal change in the opening area of the rotary valve is designed to be equivalent to that of the valve port of a typical 4-cycle internal combustion engine. The difference between the phases for the area changes of two ports is a half cycle $\tau/2$ (τ is the period of one cycle of the area change of the rotary valve).

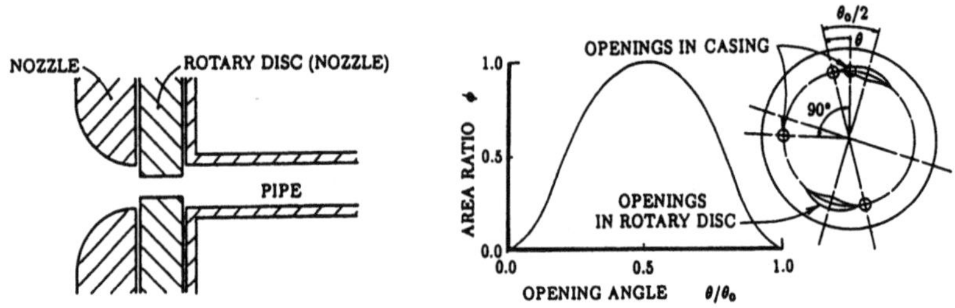

(a) Geometry (b) Area Change

Figure 6. Rotary Valve

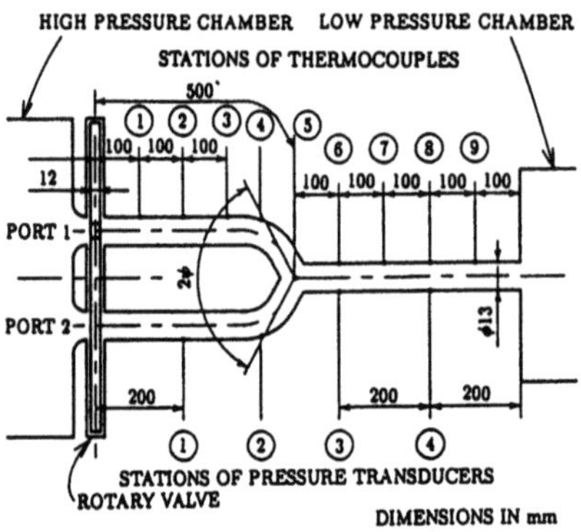

Figure 7. Geometry of Pipe with Junction

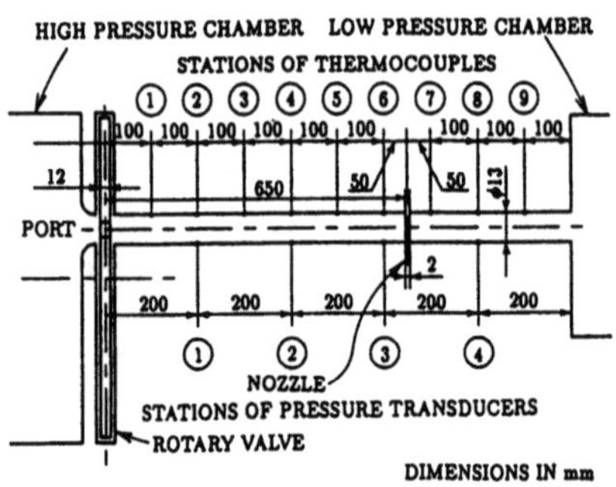

Figure 8. Geometry of Pipe with Nozzle

The detailed geometry of the pipes with each pipe elements is shown in Figs. 7 to 9. The pipe which linkes the high pressure chamber to the low pressure chamber is 1 m long, its internal diameter being 13 mm. The junction, in which the angle between two of any pipes is equal to each other, i.e. $2\psi = \frac{2}{3}\pi$, is located at the position 500mm downstream of the rotary valve. The nozzle of which area ratio to pipe $\phi = \frac{1}{2}$ or $\frac{1}{4}$ is located at the position 650 mm and the elbow of which the turning angle $\omega = \frac{1}{3}\pi$ is at 750mm.

These pipes have 4 stations along themselves and the distances between neighboring stations are 200 mm. The pressure histories are measured at each station at the rotational speeds of the rotary valve ranging from 500 rpm to 4000 rpm.

3.2 Numerical Simulation

In practice, boundary conditions at inlet and outlet of the pipe are required to specify for the numerical simulation of the flow field mentioned above. The boundary conditions at pipe end fitted to the low pressure chamber are given under the assumption of quasi-steady flow and the three-dimensional effect of the flow is grossly considered in the boundary conditions[5]. The boundary condition at another pipe end, i.e. at the rotary valve are given as the combination of the boundary conditions at the rotary disc and the convergent nozzle to the rotary disc from the high pressure chamber as shown in Fig. 6(a). The area

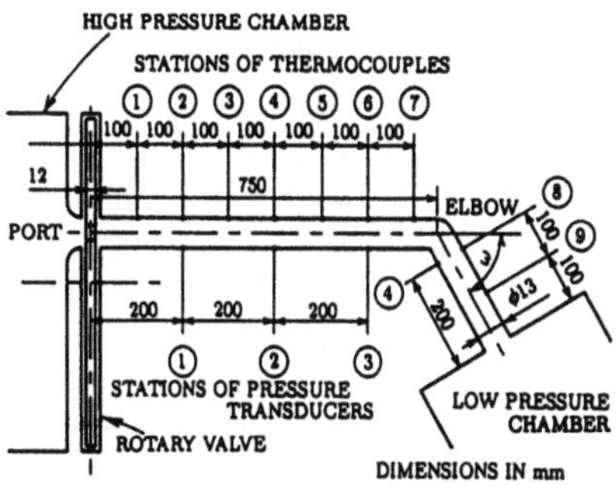

Figure 9. Geometry of Pipe with Elbow

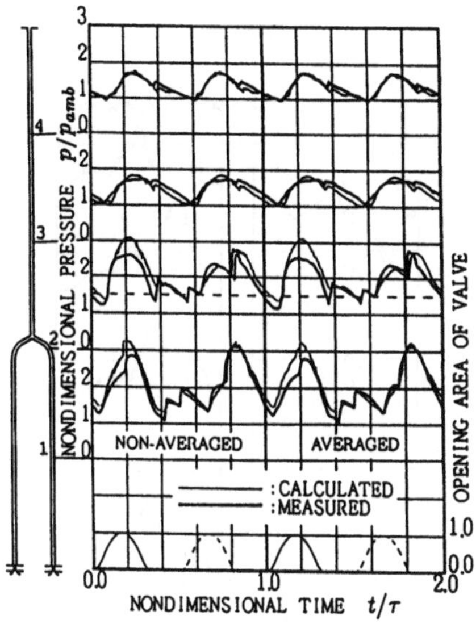

Figure 10. Pressure Histories along pipe with Junction

change of the rotary valve is obtained as the continuous change of the area ratio of the nozzle to pipe and so, the boundary condition at the nozzle is applied to that of the rotary valve. The boundary condition at the pipe inlet, i.e. the convergent nozzle to the rotary disc is given under the assumption that the flow at the pipe inlet is quasi-steady and isentropic, and Riemann variables are constant along the left-running characteristics[(4)].

In this simulation, the frictional force and the heat transfer are taken into account in the given equations, and the flow field is divided into 200 grids along the pipe. At all grid points, the atmospheric conditions with the zero-velocity are initially imposed and the numerical calculation is made during the period of first 20 cycles of the area change of the rotary valve.

4 Results and Discussions

The numerical pressure histories along the pipe with the junction, the nozzle or the elbow are compared with the experiments. These are shown in Figs. 10 to 12, respectively. The pressure histories are obtained at the rotational speed of the rotary valve n=2000 rpm. The pressure histories shown by thick lines are obtained by experiment and thin lines by calculation. The ordinate is the pressure nondimensionalized by the ambient pressure p/p_{amb} at each station along the pipe shown in Figs. 7 to 9. The abscissa is the time nondimensionalized by the period of one cycle of the rotary valve t/τ. The nondimensional opening area change of the rotary valve is shown on the bottom of the figures. The dashed line parallel to the abscissa denotes the location of each pipe element. The numerical results during the period of the last cycle of calculation is shown between t/τ=0.0 to 1.0, and the ensemble averaged pressure histories over the period of 20 cycles between t/τ=1.0

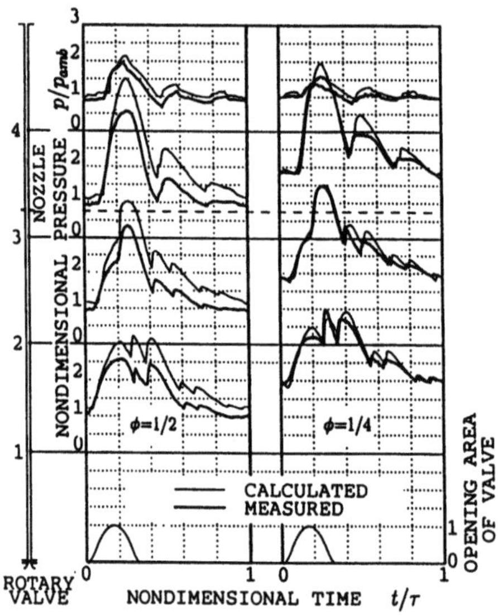

Figure 11. Pressure Histories along pipe with Nozzle

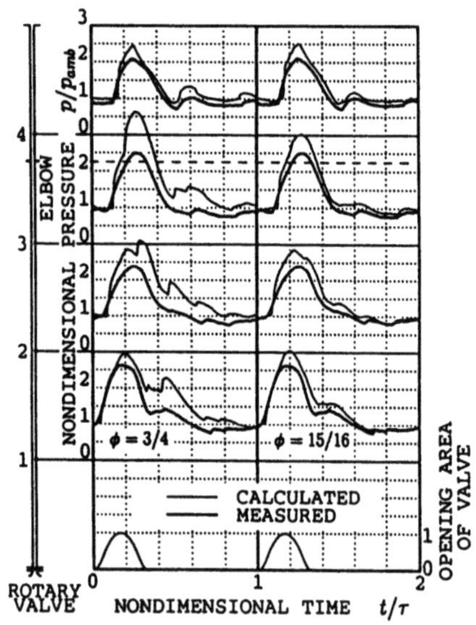

Figure 12. Pressure Histories along pipe with Elbow

to 2.0 in Fig. 10. The non-averaged pressure histories have small oscillation and these are removed by ensemble-averaging and thus, the shock waves are slightly smeared out. Consequently, the numerical results averaged as above are compared with the experimental ones for the flow field in the pipe with the nozzle or the elbow.

4.1 Junction and Nozzle

As can be seen in Figs. 10 and 11, the pressure histories are simulated very well. In particular, as long as the shock location is concerned, the agreement between calculation and experiment is good, its error being less than 5% of the period of one cycle of the rotary valve. Regarding the amplitude of the pressure histories, the numerical results are generally greater than experimental ones and its difference reaches about 20% at maximum.

One of the reasons for this difference is considered to lie in the derivation of the boundary conditions for the junction and the way of extending the boundary conditions for the junction to those at the nozzle. First, the flow through the nozzle is actually associated with the pressure losses, and the losses are usually considered in boundary conditions with the empirical coefficient. However, as mentioned above, the boundary conditions for three-way junction introduced in this paper are based on the wave motion without any pressure losses. And the influence of the junction on the flow is overestimated since the flow through the junction is assumed to turn articulately. That is, the reflective wave becomes greater and the wave through the junction becomes smaller than those by experiment.

Second, the way of extending the boundary conditions for the junction to those for the nozzle is developed under the assumption that there is no three-dimensional flow around the nozzle. In fact, the flow reaching the nozzle is forced to turn to the opening of the nozzle by three-dimensional effect and the flow velocity on the nozzle plate does not vanish, while there is little three-dimensional effect on the flow field downstream of the nozzle except the shear flow at the jet boundary. Accordingly, it is considered to be impossible to ignore three-dimensional effect in the flow field upstream of the nozzle with large area ratio. Thus, the numerical pressure on the flow field between the rotary valve and the nozzle is larger than the experimental one as can be seen in Fig. 11.

As can be seen in the pressure histories at the stations 1 and 2 in Fig. 10, concerning the pressure waves generated while the valve is open, which are about to propagate downstream through the pipe, the pressure amplitude by calculation is greater than that by experiment. The amplitudes of other pressure waves which move through the junction are comparable to each other. This can be considered to be, in part, due to the assumption of one-dimensional flow in deriving the boundary conditions for the junction. Since the influence of the geometry of junction on the flow is considered to be overestimated, the numerical amplitude becomes usually greater than the experimental one. The fact that the difference between the amplitude of numerical and experimental pressure histories downstream of the junction, i.e. at the stations 3 and 4 is small ensures this statement.

4.3 Elbow and Bend

In Fig. 12, the experimental pressure histories along the pipe with the elbow with turning angle $\omega = \frac{1}{3}\pi$ are compared with numerical ones obtained under the different boundary conditions. Comparing of Eq. (6) to Eq. (9), the elbow with turning angel $\omega = \frac{1}{3}\pi$ is found to influence the flow field in the similar way as the nozzle with the area ratio $\phi = \frac{3}{4}$. The numerical pressure histories obtained by Eq. (9) are shown as "$\phi = 3/4$". As can been seen in the pressure histories for $\phi = \frac{3}{4}$, the difference between the numerical and experimental pressure amplitude is very large although the calculation qualitatively agree with the experiment. This is considered to be caused by the overestimation of the influence of the geometry of the elbow on the flow. The boundary conditions are derived by extending the boundary conditions for the junction so that the flow is assumed to behave articulately through the elbow. In practice, the inner edge of the elbow is sharp and the outer one rounded like a bend. And so, the geometry of the edge relaxes the influence of the turning angle on the flow. To modify the estimation of the influence of the geometry of edge, the above remarks are considered as follows;

The flow field through the elbow is divided into two imaginary regions; a region where the flow behavior is completely articulate and the other region where the flow field is not affected by the elbow. If it is assumed that the flow turns articulately only in the region of inner part, whose cross-sectional area is one fourth the total cross-sectional area of the pipe, the additional flow state B is combined into the imaginary flow state D obtained in Fig. 4, so that new imaginary flow state is estimated. And then, the flow state D modified for the estimation of the flow state a is calculated by the following equation.

$$A_d = \frac{1}{4}\left(\frac{1}{4}A_a + \frac{3}{4}A_b\right) + \frac{3}{4}A_b = \frac{1}{16}A_a + \frac{15}{16}A_b \tag{10}$$

As a result, comparison of Eq. (6) with Eq. (10) leads to $\phi = \frac{15}{16}$. The numerical results for $\phi = \frac{15}{16}$ are shown as "$\phi = 15/16$" in Fig. 12. As can be seen, the good agreement is obtained between the experiment and the calculation, which suggests that the boundary condition for the elbow and the bend be treated with the same unique parameter ϕ as that for the nozzle.

The other reason for the difference between experiment and calculation can be that quasi-steady, one-dimensional flow is assumed at each pipe element in spite of the fact that the flow there is complicated, unsteady and three-dimensional.

5 Conclusions

The boundary conditions for junction are derived and these are extended to the boundary conditions for the nozzle and elbow. The pulsating flow through the pipe with the junction, nozzle or elbow are simulated using the RCM with these boundary conditions. The numerical results are compared with experimental ones. As a result, the following conclusions are drawn;

(1) The agreement between numerical and experimental pressure histories is good regarding, in particular, the location of shock wave.

(2) The difference between the amplitudes of numerical and experimental results is considered to be due partly to the way of introducing the boundary conditions for the junction and extending those to the nozzle and elbow.

(3) The way of extending the boundary condition for the junction to the nozzle and elbow is found to be proper.

(4) The boundary conditions for the nozzle and the elbow are unified through the unique parameter ϕ.

6 Acknowledgment

The experiment described above was made by Messrs. M. Karube, H. Kawanishi, Y. Ishikawa and T. Takagi. The authors wish to record their thanks for the permission of the use of their experimental results.

References

(1) E. Jenny, "Unidimensional Transient Flow with Consideration of Friction, Heat Transfer, and Area Change of Section", The Brown Boveri Review, 1950, pp.447-461.

(2) R. S. Benson, "Some Recent Research on Non-Steady Flow Problems", Flow Measurement Symposium, ASME, 1966, p.80.

(3) M. Takizawa et. al., "A Study of Gas Exchange Process Simulation of an Automotive Multi-Cylinder Internal Combustion Engine", SAE Paper 820410, 1982.

(4) M. Endo and J. Iwamoto, "Numerical Analysis of Pulsatile Flow in Pipe System with Three-Way Junction", SAE Paper 952068, 1995.

(5) M. Endo and J. Iwamoto, "A Numerical Study of Pulsating Pipe Flow With and Without a Nozzle", SAE Paper 900678, 1990.

Shock Waves Around Bodies in Unsteady Flow Fields

Hartmann, Chr., Kasten, S., and Roesner, K.G.
TUD - Institut fuer Mechanik - Dynamik der Fluide, Hochschulstrasse 1,
D-64289 Darmstadt, GERMANY

Abstract

Based on Cabannes' [1] analytic investigation on nonstationary attached shock waves around accelerated or decelerated bodies at supersonic velocities, the unsteady supersonic flow around wedge-like and conical bodies is analysed analytically and numerically with respect to the relationship between the curvature of the body contour at the pointed nose and the curvature of the attached shock wave at the tip of the body. The body is transformed to rest, and the instantaneous value of the acceleration or deceleration of the flow field is taken into account in the basic equations. In its lowest order of approximation the acceleration or deceleration is linearly combined in a relationship with the curvature of the body shape and the shock at the nose point. By computer algebraic calculations the higher order terms of a series expansion of the relation between the curvatures and the acceleration are determined and compared with numerical simulations on the basis of the Eulerian and the compressible Navier-Stokes-equations.

1. Introduction

For supersonic flows at Mach numbers close to 1 the behavior of the flow quantities near the attached shock wave are essentially influenced when the flow is accelerated or decelerated. Since Cabannes' investigations [2],[3] on nonstationary attached shock waves, a great interest was shown for the experimental and numerical analysis of nonstationary shock wave problems. Whitham's paper [4] on the flow pattern of a supersonic projectile and Rao's [5] analysis of supersonic bangs, mark the origin of the research concerning the flow near the pointed nose of conical bodies in nonstationary flows.

In the present paper a numerical analysis is performed for the determination of the shock curvature on the basis of a time-dependent calculation of the flow field around accelerated or decelerated bodies of revolution or around wedges. A TVD-scheme is used with an explicit discretization with respect to time. The Riemann solver according to Harten, Lax, and van Leer is applied. The calculations are performed on a Fujitsu machine VPP 300/6 of the Darmstadt Computing Center of the Darmstadt University of Technology. A comparison with the known data from Cabannes' asymptotic analysis is given which shows some differences between the numerical results and the analytical formula which leads to a linear relationship between the curvature of the shock, the curvature of the body shape at the nose κ_{body} and the acceleration of the body:

$$\kappa_{shock}(t) \;=\; f(\theta, M, \gamma)\,\kappa_{body} \;+\; g(\theta, M, \gamma)\,\frac{a(t)}{c^2}\,. \tag{1}$$

$\kappa_{shock}(t)$ means the instantaneous curvature of the shock wave, f and g are functions of the angle of the velocity with respect to the symmetry line of the flow field, M is the Mach number, γ is the ratio of the specific heat at constant pressure and constant volume, respectively, the acceleration is denoted by a(t), and c is the speed of sound.

2. Basic Equations

The time-dependent motion of wedge-like or conical bodies at supersonic velocities in a gas at rest is described relative to a coordinate system which is moving with the body. Therefore, in the case of an accelerated body, the upstream velocity has the negative sign of the instantaneous velocity of the body. In addition, the equation of conservation of momentum is enlarged on the right hand side by an additional force term. The Eulerian equations are used in the following form:

$$\frac{D\vec{v}}{Dt} = -\frac{1}{\rho}\nabla p - \vec{a}, \tag{2}$$

while the Navier-Stokes equations are written in the form:

$$\frac{D\vec{v}}{Dt} = -\frac{1}{\rho}\nabla p - \vec{a} + \mu\Delta\vec{v}, \tag{3}$$

where ρ is the density, $\vec{v}$ the velocity, and p the pressure and μ the dynamic viscosity of the gas. The operator $\frac{D}{Dt}$ denotes the differentiation along path lines in the flow field, and $\vec{a}$ is the acceleration of the moving body. Also in the energy equation an additional term is added on the right hand side which results in the following form of the equation of conservation of energy for the Eulerian equations:

$$\frac{D}{Dt}\left(e + \frac{1}{2}\vec{v}^2\right) = -\frac{1}{\rho}\nabla(p\,\vec{v}) - \vec{a}\cdot\vec{v}. \tag{4}$$

e is the specific internal energy of the gas. In the same way the energy equation for a viscous gas is enlarged by the acceleration term on the right hand side. The continuity equation is used in the form:

$$\frac{D\rho}{Dt} + \rho\nabla\cdot\vec{v} = 0. \tag{5}$$

As initial condition for the numerical simulations, the calculations are started from a supersonic flow at constant Mach number far upstream. The specific internal energy at infinity is calculated according to the formula:

$$e_\infty = \frac{1}{\gamma(\gamma-1)}\frac{\vec{v}_{f0}^2}{M_{0\infty}^2}. \tag{6}$$

$\vec{v}_{f0}$ and $M_{0\infty}$ are the upstream velocity and the Mach number of the oncoming gas flow. The boundary conditions are fulfilled according to Fig. 1. The distribution of the flow variables at the inlet $\partial\Omega_1$ are given by the following formulas:

$$\begin{aligned} \rho(t)|_{\partial\Omega_1} &= \rho_\infty, \\ \rho(t)\vec{v}(t)|_{\partial\Omega_1} &= -\rho_\infty\vec{v}_f(t), \\ E(t)|_{\partial\Omega_1} &= \rho_\infty e_\infty + \frac{1}{2}\rho_\infty\vec{v}_f^2(t). \end{aligned} \tag{7}$$

The velocity $\vec{v}_f$ is the instantaneous velocity of the body, when described in an inertial frame of reference. At the section $\partial\Omega_2$ outflow boundary conditions are assumed. Along $\partial\Omega_3$ in front of the body a symmetry condition is fulfilled, and on the body contour - which may be curved or a straight line - we impose for the Eulerian equations the condition

$$\vec{v}\cdot\vec{n} = 0, \tag{8}$$

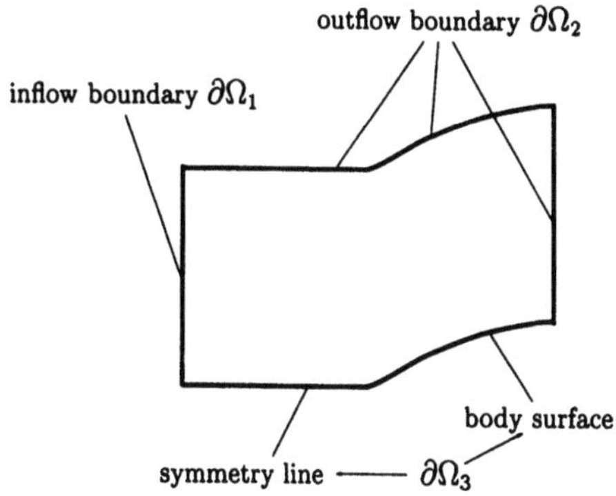

Figure 1: Flow region

$$\vec{v}_{body} = \vec{0}\,, \tag{9}$$

where $\vec{n}$ denotes the local normal unit vector on the body surface. In addition an adiabatic behavior of the flow field is assumed on the body surface. E is the specific total energy. Further boundary conditions are imposed on ρ and E:

$$\nabla\rho\cdot\vec{n} = 0; \text{ and } \nabla E\cdot\vec{n} = 0\,. \tag{10}$$

3. The Algorithm For the Eulerian Approach

For the numerical simulation of the time-dependent flow around wedges or cones, a body fitted coordinate system is used. In the physical plane the region close to the body nose is spatially resolved by a fine grid, and also in the neighborhood of the body surface the grid lines are attracted. Several grids were used for the calculations to show that the results are independent of the chosen grid distribution.

The basic system of equations is written in conservative variables as follows:

$$(\vec{U})_t + (\vec{F}(\vec{U}))_x + (\vec{G}(\vec{U}))_y + \delta\vec{B}(\vec{U}) = \vec{A}(\vec{U}). \tag{11}$$

δ denotes the values 0 or 1 according to the plane or axisymmetrical geometry under consideration, respectively. The vector functions $\vec{U}, \vec{F}, \vec{G}, \vec{B}$, and $\vec{A}$ are defined in the following way:

$$\vec{U} = \begin{pmatrix} \rho \\ \rho u \\ \rho v \\ E \end{pmatrix}, \tag{12}$$

$$\vec{F}(\vec{U}) = \begin{pmatrix} \rho u \\ \rho u^2 + (\gamma-1)\left[E - \frac{1}{2}\rho(u^2+v^2)\right] \\ \rho uv \\ u\left[\gamma E - (\gamma-1)\frac{1}{2}\rho(u^2+v^2)\right] \end{pmatrix}, \tag{13}$$

$$\vec{G}(\vec{U}) = \begin{pmatrix} \rho v \\ \rho vu \\ \rho v^2 + (\gamma-1)\left[E - \frac{1}{2}\rho(u^2+v^2)\right] \\ v\left[\gamma E - (\gamma-1)\frac{1}{2}\rho(u^2+v^2)\right] \end{pmatrix}, \tag{14}$$

$$\vec{A}(\vec{U}) = \begin{pmatrix} 0 \\ -\rho a_x \\ -\rho a_y \\ (-\rho \vec{a})\vec{v} \end{pmatrix}, \qquad \vec{B}(\vec{U}) = \frac{1}{y} \begin{pmatrix} \rho v \\ \rho v u \\ \rho v^2 \\ v\left[\gamma E - (\gamma - 1)\frac{1}{2}\rho(u^2 + v^2)\right] \end{pmatrix}. \tag{15}$$

The numerical fluxes are modelled by an approximate solution of the Riemann-problem. The flow variables are assumed to be constant in each cell. At the cell boundary the flux is calculated by the values of the left and right neighbor cell. This leads by extrapolation with a weighting factor k to the following formulas:

$$\vec{\mathbf{U}}^{n(L)}_{i+\frac{1}{2}} = \vec{\mathbf{U}}^n_i + \frac{1}{4}\left[(1-k)(\vec{\mathbf{U}}^n_i - \vec{\mathbf{U}}^n_{i-1}) + (1+k)(\vec{\mathbf{U}}^n_{i+1} - \vec{\mathbf{U}}^n_i)\right], \tag{16}$$

$$\vec{\mathbf{U}}^{n(R)}_{i+\frac{1}{2}} = \vec{\mathbf{U}}^n_{i+1} - \frac{1}{4}\left[(1+k)(\vec{\mathbf{U}}^n_{i+1} - \vec{\mathbf{U}}^n_i) + (1-k)(\vec{\mathbf{U}}^n_{i+2} - \vec{\mathbf{U}}^n_{i+1})\right]. \tag{17}$$

This is used for an approximation of the numerical fluxes of second order accuracy. The time integration is performed by a splitting into two half steps to gain second order accuracy. The whole algorithm is split for the two-dimensional problem which leads to the following sequence of operations

$$\vec{\mathbf{U}}^{n+1} = \mathcal{L}_\xi(\frac{\Delta t}{2})\mathcal{L}_\eta(\Delta t)\mathcal{L}_\xi(\frac{\Delta t}{2})\vec{\mathbf{U}}^n. \tag{18}$$

4. Results of Eulerian Equations and Discussion

The main problem for the comparison of numerical data with the analytical formulas due to Cabannes is the identification of the shock front. In the present investigation the curvature of the shock wave is determined by the least square method. The points on the shock wave are calculated by a polynomial ansatz of second degree. In Fig. 2 the approximation of a shock wave is shown. The variation of the radii of curvature for

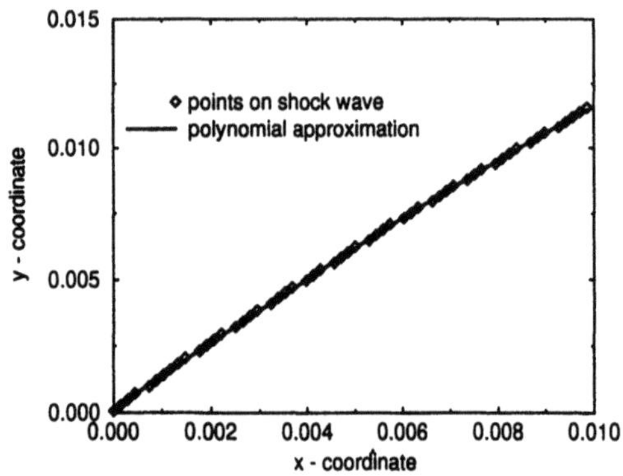

Figure 2: Polynomial approximation of the shock wave contour

an accelerated and also for a decelerated wedge is given in the Fig. 3. The two curves show the radii of curvature for several Mach numbers. Starting with the initial condition at t=0 for the flow field variables at Mach number M=1.8 the radii of curvature are calculated for different times when the Mach number changes due to the acceleration of the wedge. Therefore, the Mach number variation corresponds to the increase of time for which the numerical simulations were carried out. For a decelerated wedge the starting Mach number was chosen to be M=2.5, and the radii of curvature are plotted in Fig. 3 down to the Mach number of M=1.7.

As an example of the time-dependent flow around a cone with an aperture angle of 20^0

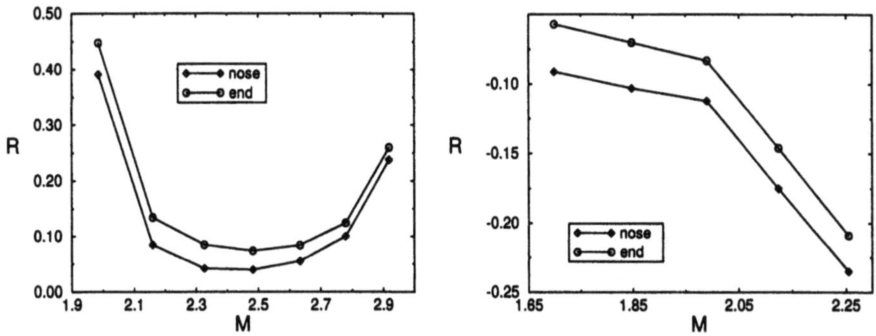

Figure 3: Radii of curvature of the shock wave for an accelerated wedge (left) and a decelerated one (right) ($\theta_s = 15°$)

in Fig. 4 the time sequence is shown for the evolution of the flow field for the acceleration a=25. The Mach number varies from the initial value M=1.3 to its final value M=2.193. This plot of the isopycnic curves demonstrates the change of the curvature of the shock wave during a time interval of Δt=0.02746. A comparison of the asymptotic results due to Cabannes with the numerical data is shown in Fig. 5. The calculations were started from a flow field around a wedge at Mach number M=2.5. At the nose and at the end of the finite body the radii of the curvature are plotted. For the decelerated wedge (a= -20) with the aperture angle of 30^0 a good agreement is seen for the Mach number interval

$$1.9 < M < 2.25.$$

It has to be taken into account, that a finite length of the body will naturally influence the flow field around the nose of a body, as experiments and numerical investigations have shown. Under the expansion wave which is present at the rear part of a wedge or cone with finite length the shape of the whole shock wave will be influenced. Insofar the numerical results are valid only for very long bodies in unsteady supersonic flows.

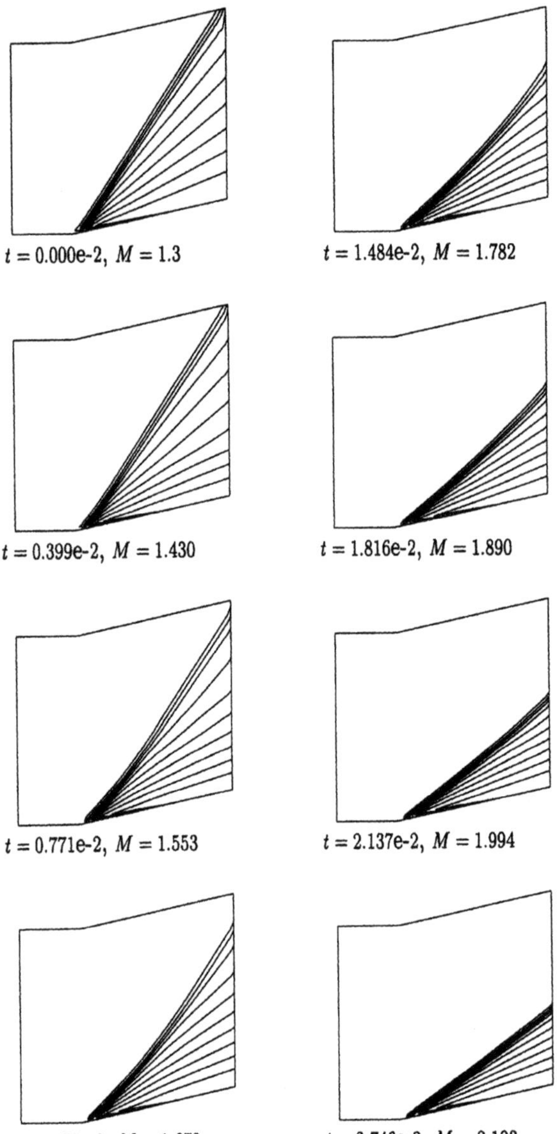

$t = 0.000\text{e-}2,\ M = 1.3$

$t = 0.399\text{e-}2,\ M = 1.430$

$t = 0.771\text{e-}2,\ M = 1.553$

$t = 1.138\text{e-}2,\ M = 1.670$

$t = 1.484\text{e-}2,\ M = 1.782$

$t = 1.816\text{e-}2,\ M = 1.890$

$t = 2.137\text{e-}2,\ M = 1.994$

$t = 2.746\text{e-}2,\ M = 2.193$

Figure 4: Isopycnic curves of the flow field around an accelerated cone ($\theta_s = 10°$, $a = 25$) for different time steps

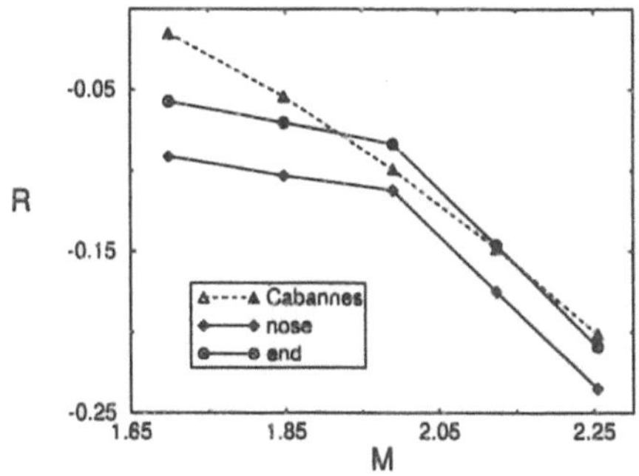

Figure 5: Comparison of the analytical and numerical radii of curvature for a decelerated wedge ($\theta_s = 15°$, $a = -20$)

5. Comparison With Results From Compressible Navier-Stokes Equations

All numerical simulations for the Eulerian and the Navier-Stokes equations were started from a stationary flow field at a given Mach number (Fig. 6). In the following the results for the flow around a wedge of 10° half opening angle are summarized, and starting with the Mach number 2, the same number of time steps were followed in the evolution of the flow field, once for the nonviscous flow and then for the viscous flow for a Reynolds number of 50000. The Reynolds number was defined according to:

$$Re = \frac{L\, c_\infty}{\nu_\infty}. \tag{19}$$

The typical length scale L is given by the length of the ramp of the wedge, c_∞ and ν_∞ are the speed of sound and the kinematic viscosity at infinity.

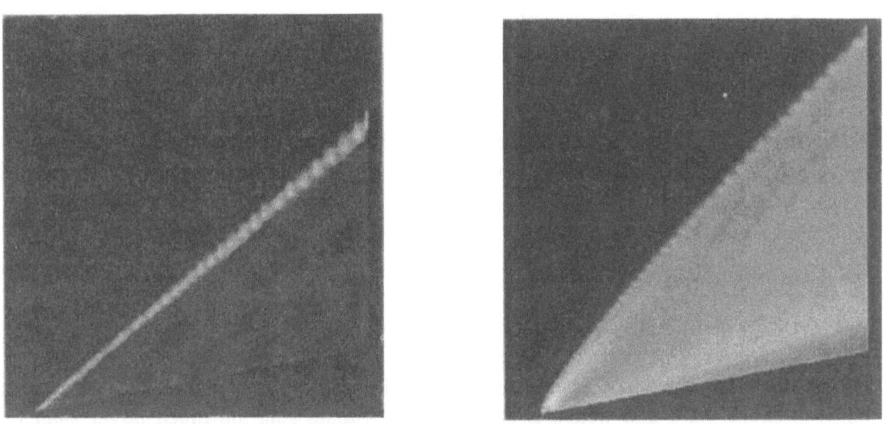

Figure 6: Stationary solution for the flow around a wedge of half opening angle of 10° at Mach number Ma=2. Left: Euler, right: Navier-Stokes

It can be seen from Fig. 6 that for the stationary flow field the attached shock waves of the Navier-Stokes calculation shows a much stronger curvature at the tip of the wedge than the straight shock, calculated for the Eulerian equations.

If the evolution of the flow field and the shock is followed for both the cases (Euler and Navier-Stokes) up to the time

$$t = 3.68 \times 10^{-2}, \tag{20}$$

which corresponds to a dimensionless acceleration of a=20, and a Mach number of the oncoming gas stream of

$$M = 3.47, \tag{21}$$

the shock in nonviscous flow already starts to get a slightly concave shape (Fig. 7) while the shock front of the Navier-Stokes solution still remains convex, looking from the upstream direction.

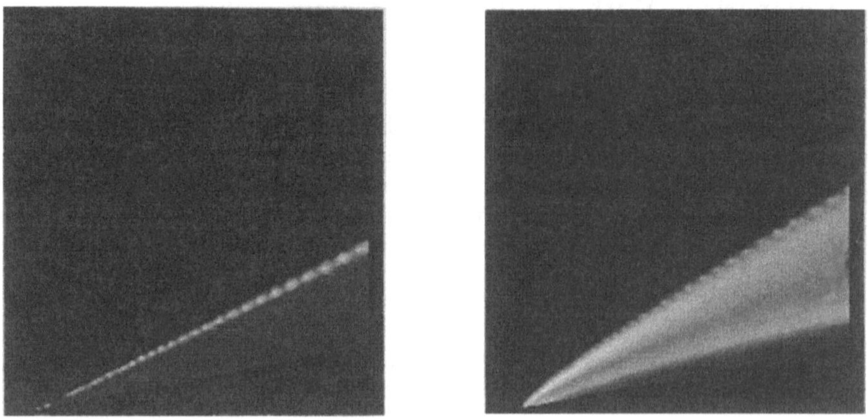

Figure 7: Attached shock wave at the wedge with half opening angle of 10° in a flow with constant acceleration of a=20 at time $t = 3.68 \times 10^{-2}$, ($M = 3.47$). Left: Euler, right: Navier-Stokes

This behavior of the two solutions from the nonviscous and the viscous flow calculations is typical for even later times of the evolution. In Fig. 8 the two flow fields are compared for the time step

$$t = 0.11, \tag{22}$$

and a corresponding Mach number

$$M = 6.27. \tag{23}$$

It is evident that the shock approaches more and more the body surface, but the region between shock wave and wedge is for the Navier-Stokes solution much larger than for the Eulerian solution. The viscosity effects lead naturally to a boundary layer along the wedge which thickens the region between body and shock. Therefore, a comparison of the two flow fields should take into account the boundary layer thickness for the viscous flow in order to correct the half opening angle of the wedge for the non viscous flow.

It seems to be evident that a realistic shock wave simulation has to be based on the compressible Navier-Stokes equations. This would also have a strong influence on the calculation of the curvature of the attached shock at the tip of the body.

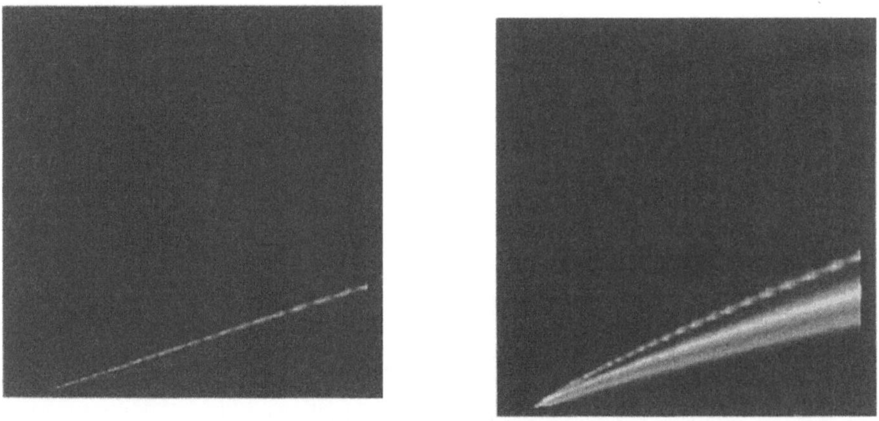

Figure 8: Accelerated supersonic flow over a wedge of half opening angle of 10° at time $t = 0.11$, $(M = 6.27)$. Left: Euler, right: Navier-Stokes

References

[1] Cabannes, H. Etude de l'onde de choc atachée dans les écoulements de révolution, La Rech. Aéro. no. 24, (1951), 17-23

[2] Cabannes, H. Etude de l'onde de choc atachée dans les écoulements de révolution, La Rech. Aéro. no. 27, (1952), 7-16

[3] Cabannes, H.: Influence des accélérations sur la courbure des chocs, La Rech. Aéro., no. 39, (1954), 2-13

[4] Whitham, G.B. The Flow Pattern of a Supersonic Projectile, Comm. Pure and Appl. Math., Vol. 5, (1952), 301-348

[5] Rao, P.S. Supersonic Bangs, Part I and II, Aeron. Quart., 7, (1956), 135-155

A Numerical Investigation of Mixed Convection on a Horizontal Semi-Infinite Plate

Valod Noshadi and Wilhelm Schneider

Institute of Fluid Dynamics and Heat Transfer, Technical University of Vienna
Wiedner Hauptstrasse 7/322, A-1040 Vienna, Austria

Dedicated to Prof. I. Teipel on the occasion of his 65th birthday.

Abstract. The laminar mixed convection flow above a semi-infinite horizontal plate is studied for large Reynolds and Grashof numbers. To permit similarity solutions of the boundary-layer equations, the plate temperature, T_w, is assumed to differ from the free-stream temperature, T_∞, according to the relation $(T_\infty - T_w) \propto x^{-\frac{1}{2}}$, where x is the distance from the leading edge. This corresponds to a plate that is strongly cooled near the leading edge while thermally insulated everywhere else. There are two branches of similarity solutions of the boundary-layer equations if the non-dimensional buoyancy parameter is larger than a critical value, which is negative and depends on the Prandtl number. Non-similar solutions connecting the self-similar solutions have also been found analytically. In the present work a finite-element package is used to solve the full Navier–Stokes and energy equations. The numerical results resemble the analysis in a certain regime of the buoyancy parameter. In other regimes, however, neither the self-similar solutions nor the connecting solutions are found numerically.

Keywords. Mixed convection, numerical solutions, Navier–Stokes equations

1 Introduction

Consider a semi-infinite horizontal plate aligned parallel to a uniform free stream with velocity u_∞ and temperature T_∞, cf. figure 1. u and v are the velocity components in x and y directions, respectively, of a Cartesian coordinate system with the origin at the leading edge of the plate, g is the gravity acceleration, and the temperature at the plate surface, T_w, may vary with the coordinate x. Since the temperature in the boundary layer differs from the ambient temperature, there is a hydrostatic pressure difference across the boundary layer. As the boundary-layer thickness increases with increasing distance from the leading edge, the hydrostatic pressure at the plate surface varies too, giving rise to a favourable or adverse pressure gradient in the case of a heated ($T_w > T_\infty$) or cooled ($T_w < T_\infty$) surface, respectively, provided the thermal expansivity, β, is positive.

Previous investigations have shown (Schneider 1979) that the mixed convection flow on

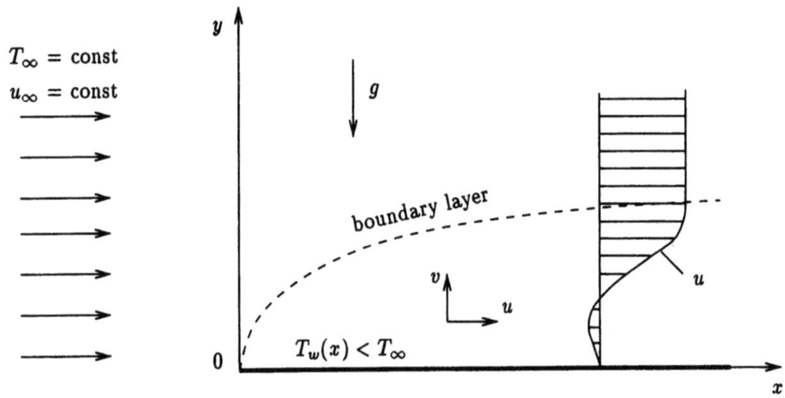

Figure 1: Problem.

a horizontal semi-infinite thermally insulated plate with a strong heat source or sink at the leading edge gives a wall temperature distribution according to $(T_w - T_\infty) \propto \pm x^{-\frac{1}{2}}$, which allows a self-similar solution of the boundary-layer equations. The dimensionless buoyancy parameter

$$K = \frac{Gr_x}{Re_x^{\frac{5}{2}}} = g\beta(T_w(x) - T_\infty)(x\nu)^{\frac{1}{2}} u_\infty^{-\frac{5}{2}} ,$$

which appears in the equations in addition to the Prandtl number Pr, does not change along the plate. Here Re_x and Gr_x are the Reynolds and Grashof numbers, respectively, and ν is the constant value of the kinematic viscosity of the fluid. Both Re_x and Gr_x are assumed to be very large, except near the leading edge of the plate.

$K = 0$ corresponds to the Blasius solution. In case of a favourable pressure gradient on the plate ($K > 0$) the similarity solution is unique. In case of an adverse pressure gradient on the plate ($K < 0$) two self-similar solutions of the boundary-layer equations exist for $K_c \leq K < 0$ (Schneider 1979; Afzal & Hussain 1984; De Hoog, Laminger & Weiss 1984; Merkin & Ingham 1987). Figure 2 shows the dimensionless wall shear stress $f''(0)$, plotted versus the buoyancy parameter K for $Pr = 1$. Here f is the Blasius stream function and the prime denotes differentiation with respect to the Blasius similarity variable η. For $K < K_c$ no self-similar solution is found, and the behaviour of the flow seems to be unknown (Schneider 1995; Steinrück 1997).

Steinrück (1995) showed that there are steady-state, non-similar, solutions of the boundary-layer equations, which connect the upper and lower branches of the self-similar solutions. Near the leading edge the connecting solutions are close to the upper-branch self-similar solution, while far downstream they approach the lower-branch solution. The

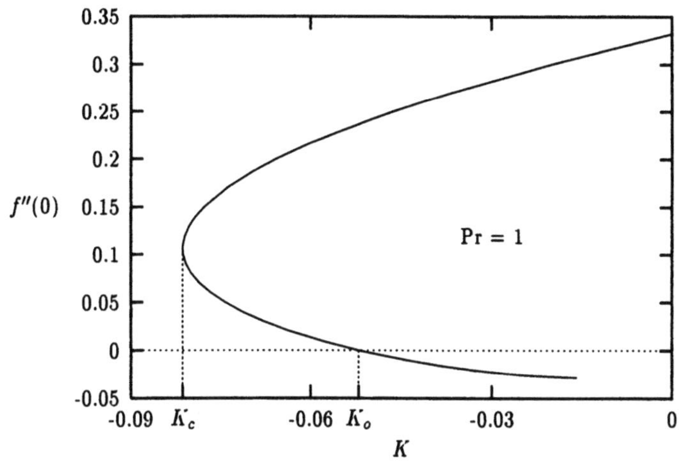

Figure 2: Self-similar solutions: dimensionless shear stress at the plate versus buoyancy parameter K.

coordinate x can be stretched arbitrarily with a constant coefficient that remains undetermined in the connecting solutions of the boundary-layer equations.

To gain further insight into the problem of mixed convection on a horizontal semi-infinite plate with adverse pressure gradient, the set of Navier–Stokes and energy equations is solved numerically in the present work.

2 Computational domain and boundary conditions

The finite-element package FIDAP (version 7.6) is used. The basic equations are discretized with 9 node quadrilateral elements. To reduce both memory requirements and computational time the pressure is eliminated from the discretized set of equations using the penalty function approach. The computational domain is shown in figure 3, where the boundary conditions are also summarized. A temperature distribution according to the self-similar solution, i.e. $(T_\infty - T_w) \propto x^{-\frac{1}{2}}$, as well as the no-slip condition are prescribed at the plate. Since the above-mentioned temperature distribution on the plate has a singularity at the point $x = 0$, it is necessary to prescribe a large, but finite temperature value for the first node at the leading edge of the plate. To simulate, as closely as possible, the necessary conditions for self-similar flow, characterized by an adiabatic wall, the temperature for the first node is chosen such that the total heat flux at the plate becomes minimum. Symmetry conditions are prescribed in the plane of the plate in front of the plate, i.e. for $x < 0$, $y = 0$. The distance of the left-hand side boundary, where the free stream conditions are prescribed, from the edge of the plate is chosen sufficiently large to

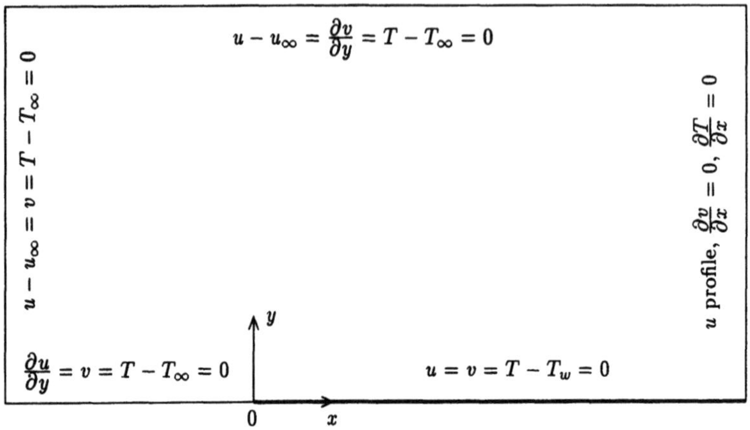

Figure 3: Computational domain and boundary conditions.

ensure a smooth decay of the horizontal component of velocity at the symmetry plane. The distance of the upper boundary from the plate is taken at least twice as large as the boundary-layer thickness. At the right-hand side (downstream) boundary the horizontal velocity profile according to the similarity solution is prescribed. Furthermore it is assumed that there are no temperature and vertical velocity gradients at the downstream boundary. The computations are performed for two different plate lengths corresponding to $\mathrm{Re}_x = 2 \times 10^4$ and 5×10^4, respectively.

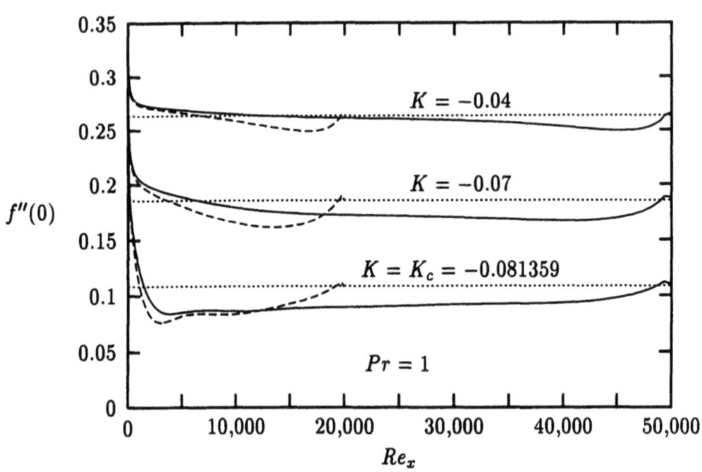

Figure 4: Dimensionless shear stress at the plate. The solid and dashed lines are the results of the numerical computations with the downstream boundary at $Re_x = 50{,}000$ and 20,000, respectively. The dotted lines correspond to the upper-branch similarity solutions.

3 Numerical solutions

3.1 Upper branch

The computations are started with the case $K = 0$, i.e. the Blasius solution. The buoyancy parameter K is then decreased in successive steps to the critical value, K_c. The solution in each step serves as initial solution for the next step of the computations. For each value of the buoyancy parameter K a horizontal velocity profile is prescribed at the downstream boundary corresponding to the *upper*-branch similarity solution of the boundary-layer equations as shown in figure 2.

In figure 4 the dimensionless wall shear stress $f''(0)$ is plotted along the plate for three different values of the buoyancy parameter K. There are (rather small) discrepancies between the numerical solutions of the Navier–Stokes equations and the similarity solutions of the boundary-layer equations, possibly due to the fact that the boundary-layer theory is not valid near the leading edge. In figure 5 the dimensionless horizontal velocity f', and the dimensionless temperature difference $\vartheta = (T - T_\infty)/(T_w(x) - T_\infty)$, are plotted versus the similarity variable η for $K = -0.04$. The velocity and temperature profiles from the numerical computations match with the profiles from the upper-branch similarity solutions of the boundary-layer equations.

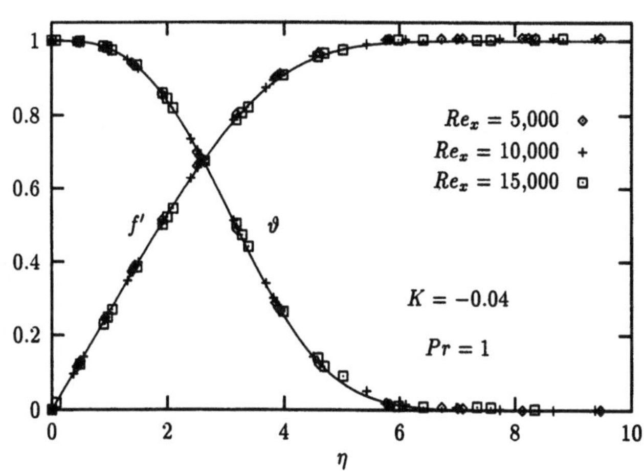

Figure 5: Dimensionless horizontal velocity and temperature profiles. The solid lines correspond to the upper-branch similarity solutions, whereas the symbols are the results of the numerical computations with the downstream boundary at $Re_x = 20{,}000$.

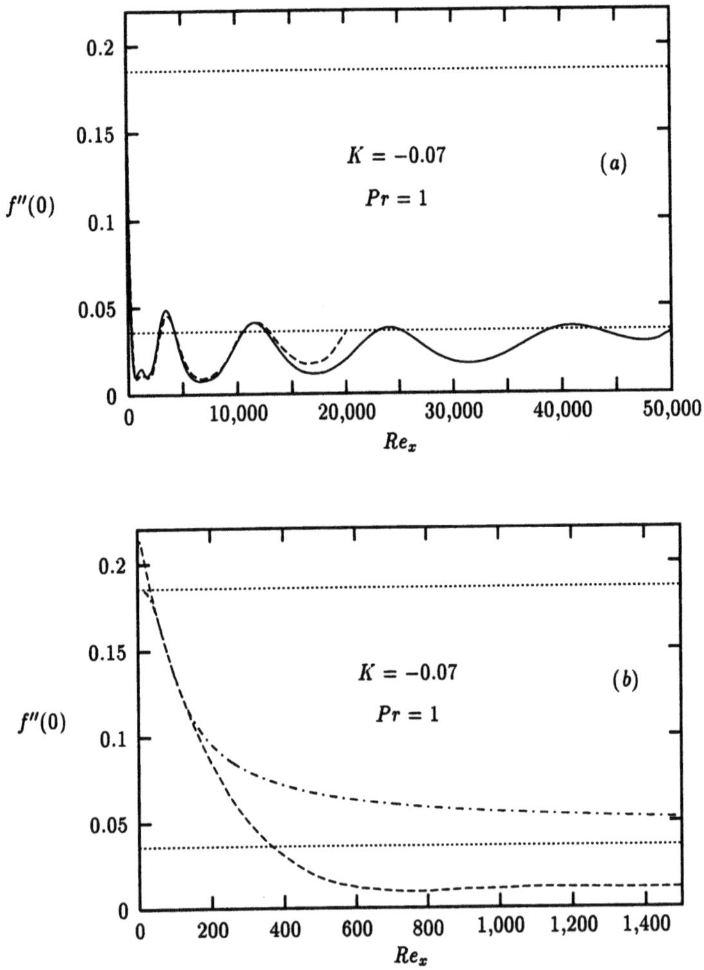

Figure 6: Dimensionless shear stress at the plate. The two dotted horizontal lines resemble the upper and lower branch similarity solutions, respectively. (*a*) Upper figure: the solid line and dashed line are the results of the numerical computations with the downstream boundary at Re_x = 50,000 and 20,000, respectively. (*b*) Lower figure: the connecting solution of the boundary-layer equations (dash-dotted line, according to Steinrück 1995) in comparison with the present numerical computations (dashed line).

3.2 Lower branch without flow reversal

To study the possible transition from one self-similar flow to the other one, the *lower*-branch similarity solution is prescribed for the horizontal velocity profile at the downstream boundary, i.e. the buoyancy parameter is assumed to be in the interval $K_c \leq K \leq K_o$, cf. figure 2. Starting with the solution for the case $K = K_c$ as initial solution, the numerical computations are continued by increasing the value of the buoyancy parameter K in successive steps. We were able to find solutions only in the range $K_c \leq K \leq -0.06$ for $Pr = 1$.

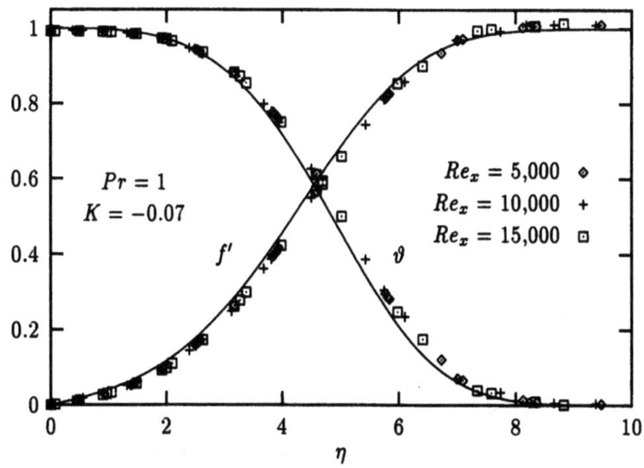

Figure 7: Dimensionless horizontal velocity and temperature profiles. The solid lines correspond to the lower branch similarity solutions, whereas the symbols are the results of the numerical computations with the downstream boundary at $Re_x = 20{,}000$.

The dimensionless wall shear stress $f''(0)$ is given in figure 6(a) for $K = -0.07$. Near the leading edge, the shear stress plots from the numerical computations show an overshooting above the value predicted by upper-branch similarity solution. Then, there is a sharp transition to the lower branch with an undershooting followed by oscillations. Although the oscillations decay with increasing distance from the leading edge, the average value appears to remain below the lower-branch self-similar solution. To check whether the observed oscillations are due to numerical instabilities, the computations were repeated with the outflow boundary located at various distances from the leading edge, with a finer mesh and with discretized pressure. No substantial changes of the solution were observed. Even transient flow computations did not show a time-dependent behaviour. Figure 6(b) shows the transition from the upper to the lower branch solutions in more detail. The free constant in the connecting solution of the boundary-layer equations (Steinrück 1995)

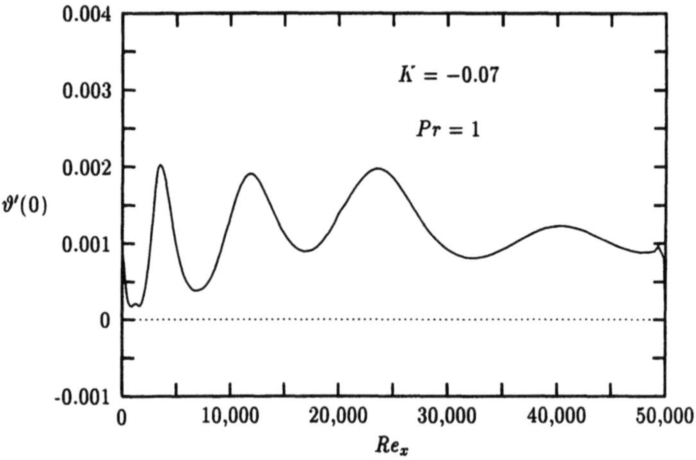

Figure 8: Computed dimensionless heat flux at the plate (solid line) in comparison with the similarity solution (dotted line).

is chosen such that the two solutions, i.e. the connecting solution (dash-dotted line) and the results of the numerical computations (dashed line), undergo transition at about the same point. With this choice of the free constant, the inclinations of the two curves are approximately equal in the transition region. In figure 7 the dimensionless horizontal velocity f', and the dimensionless temperature difference ϑ, are plotted versus the similarity variable η for $K = -0.07$. The velocity and temperature profiles from the numerical computations are again close to the profiles from the lower-branch similarity solutions of the boundary-layer equations. The very small values of the heat flux at the plate, as given in figure 8, are also in accordance with the self-similar solution, which requires vanishing wall heat flux.

3.3 Lower branch with reversed flow region

To consider the effects of flow reversal, the lower-branch similarity solution in the range $K_o \leq K < 0$ (cf. figure 2) is prescribed for the horizontal velocity profile at the downstream boundary. We were able to find numerical solutions only in the range $-0.03 \leq K < 0$ for $Pr = 1$.

The dimensionless wall shear stress $f''(0)$ along the plate is given in figure 9 for $K = -0.02$. The numerical solution depends strongly on the distance of the downstream boundary from the leading edge. According to the numerical solution the transition from the upper branch to the lower branch is rather slow, whereas the connecting solution according to Steinrück (1995) predicts a rather sharp transition. In figure 10 the dimen-

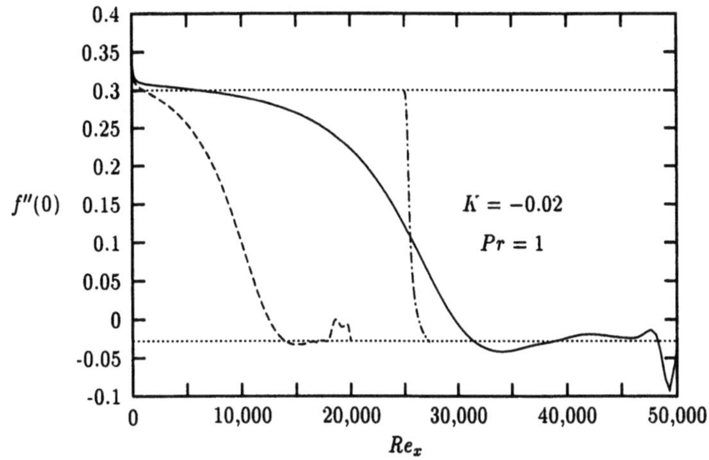

Figure 9: Dimensionless shear stress at the plate. The solid and dashed lines are the results of the numerical computations with the downstream boundary at Re_x = 50,000 and 20,000, respectively. The dash-dotted line corresponds to a connecting solution of the boundary-layer equations (Steinrück 1995). The two dotted horizontal lines resemble the upper and lower branch similarity solutions.

sionless horizontal velocity f', and dimensionless temperature difference ϑ, are plotted versus the similarity variable η for $K = -0.02$. Near the leading edge the computed velocity and temperature profiles are close to the upper-branch self-similar solution. With increasing distance from the edge of the plate the computed profiles change shape and approach the lower-branch self-similar solution.

4 Conclusions

The numerical solutions of the set of Navier–Stokes and energy equations depend on whether the downstream boundary condition is prescribed in accordance with the upper-branch or the lower-branch similarity solution of the boundary-layer equations. The dependence is as follows.

(1) If the horizontal velocity profile according to the *upper*-branch similarity solution is applied as the downstream boundary condition, the numerical solution of the full equations is in reasonable agreement with the self-similar boundary-layer solution.

(2) If the horizontal velocity profile according to the *lower*-branch similarity solution is prescribed at the downstream boundary, the comparison depends on the parameter regime. For $Pr = 1$ and values of K in the range $K_c \leq K \leq -0.06$ the numerical solutions of the full equations seem to resemble the connecting solutions of the boundary-

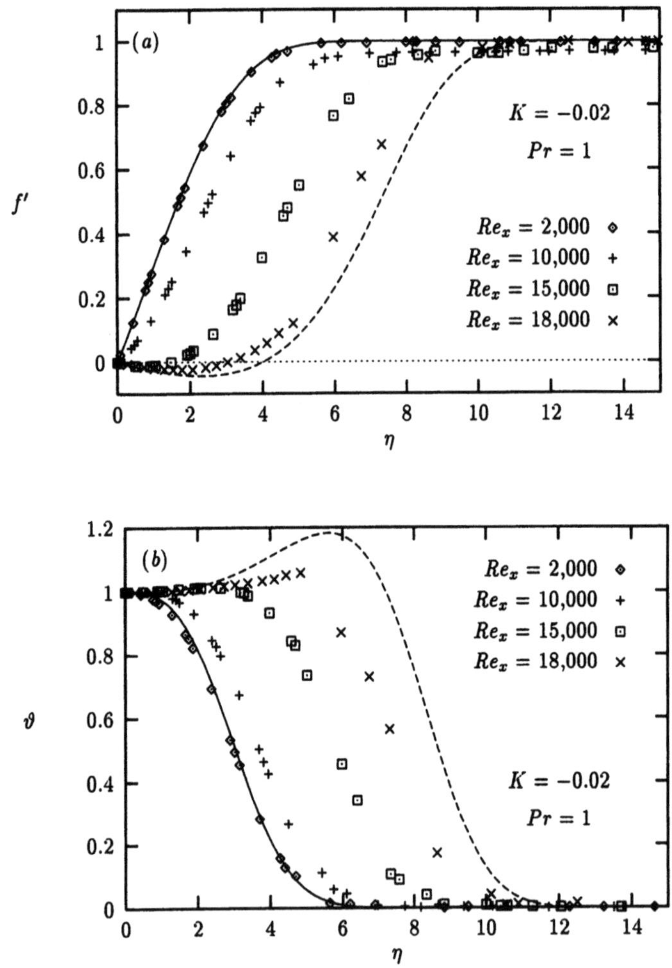

Figure 10: (a) Dimensionless horizontal velocity and (b) temperature profiles. The solid and dashed lines correspond to the lower and upper branch similarity solutions, respectively, whereas the symbols are the results of the numerical computations with the downstream boundary at $Re_x = 20{,}000$.

layer equations, although there are discrepancies between the two solutions. For $Pr = 1$ and values of K in the range $-0.03 \leq K < 0$, however, the numerical solutions of the full equations depend on the size of the computational domain and are in obvious disagreement with the available boundary-layer solutions. Between these two regimes, i.e. for $-0.06 \leq K < -0.03$ and $Pr = 1$, no numerical solution of the full equations subject to the lower-branch downstream boundary conditions could be found. It is plausible to assume that the unusual behaviour of the solutions is associated with the appearance of reversed flow regions.

Acknowledgements

This work has been supported by the "Fonds zur Förderung der wissenschaftlichen Forschung in Österreich (FWF)", project No. P9584-TEC. V. Noshadi is also grateful to the Iranian Ministry of Culture and Higher Education for financial support. The authors should like to thank Dr. H. Steinrück for many helpful discussions.

References

AFZAL, N. & HUSSAIN, T. 1984 Mixed convection over a horizontal plate. *ASME J. Heat Transfer* **106**, 240-241.

DE HOOG, F. R., LAMINGER, B. & WEISS, R. 1984 A numerical study of similarity solutions for combined forced and free convection. *Acta Mech.* **51**, 139–149.

MERKIN, J. H. & INGHAM, D. B. 1987 Mixed convection similarity solutions on a horizontal surface. *J. Appl. Math. Phys. (ZAMP)* **38**, 102–116.

SCHNEIDER, W. 1979 A similarity solution for combined forced and free convection flow over a horizontal plate. *Int. J. Heat Mass Transfer* **22**, 1401–1406.

SCHNEIDER, W. 1995 Laminar mixed convection flows on horizontal surfaces. Proc. *3rd Caribbean Cong. on Fluid Dynamics*. Vol. II, Simon Bolivar University, Caracas.

STEINRÜCK, H. 1995 Mixed convection over a horizontal plate: self-similar and connecting boundary-layer flows. *Fluid Dyn. Res.* **15**, 113–127.

STEINRÜCK, H. 1997 Recent investigations on buoyant boundary layer flows over horizontal surfaces. Submitted to *Z. angew. Math. Mech. (ZAMM)*.

Tube Bundle Vibrations due to Cross Flow Under the Influence of Turbulence

K. Popp and O. Romberg

Institute of Mechanics, University of Hannover
Appelstr. 11, 30167 Hannover, Germany

Abstract. Tube bundles are often used in heat exchangers and chemical reactors. Besides of large heat transfer capacities and small pressure drops in the apparatus a safe design against vibration damages is demanded. For many years extensive investigations concerning the dynamical behaviour of tube bundles subjected to cross-flow have been carried out in the wind tunnel of the Institute of Mechanics at the University of Hannover. In the last years the investigations were concentrated on the experimental investigations of different flow excitation mechanisms in a fully flexible bundle as well as in a bundle with one single flexibly mounted tube in an otherwise fixed array with variable geometry and changing equilibrium position. The aim of the studies was the determination of the stability boundaries, i. e. the critical reduced fluid velocity depending on the reduced damping coefficient in a wide parameter region. Theoretical investigations of the stability behaviour on the basis of an one dimensional flow model as well as experimental investigations of the influence of turbulence on the stability boundaries have been carried out. Here, for certain tube bundle configurations an increased turbulence has a stabilizing effect and leads to a shift of the stability boundaries to higher velocities. The change of the turbulence was realised by using turbulence grids at the inlet of the bundles or thin Prandtl-tripwires at the tube surfaces. Flow visualization studies at the original experimental set-up under relevant Reynolds numbers give an impression of the flow pattern. At this time an investigation of the exciting fluid forces is carried out using a flexibly mounted pressure test tube. A survey about some recent investigations is given.

Keywords. Tube bundle vibrations, wind tunnel, cross-flow, instability, turbulence, tripwire, flow visualization, pressure measurements

1 Introduction

There are different excitation mechanisms that cause fatal damages due to undesirable vibrations in heat exchanger tube bundles subjected to cross-flow, cp. Weaver & Fitzpatrick [36]. Subject of the present investigation are the two most important mechanisms which are called *fluidelastic instability* and *turbulent buffeting.* These mechanisms which can be investigated seperately influence each other. Particularly the intensity of the turbulent buffeting can have a significant influence on the onset of instability. Figure 1 illustrates

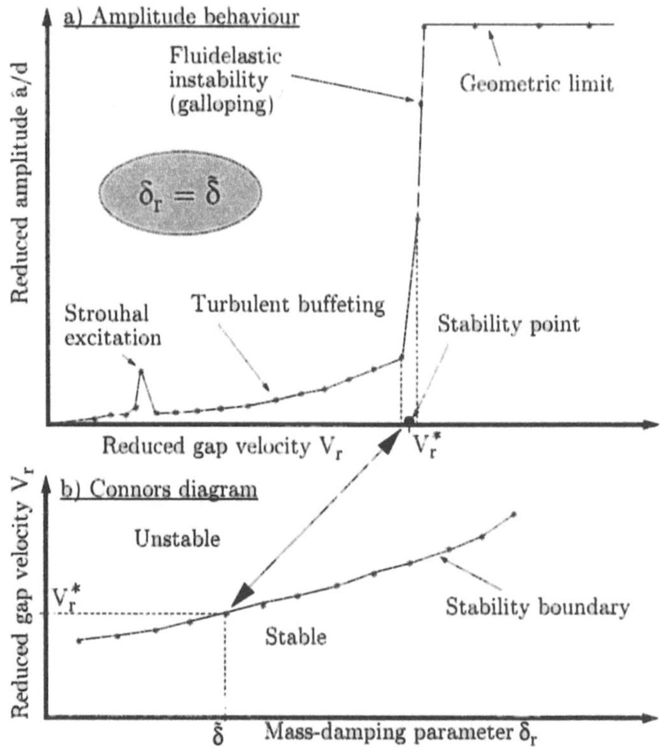

Figure 1: Experimental determination of the stability boundary of a flexible tube in a bundle subjected to cross-flow, a) amplitude behaviour, b) Connors diagram

the experimental determination of the stability threshold for a tube within a bundle. Here, the amplitude $\hat{a}$ related to the tube diameter d is plotted in dependence of the reduced gap velocity V_r (Figure 1a). A reliable determination of stability boundaries requires a stability criterion, cf. Chen [12]. In this study the amplitude criterion used also by Austermann & Popp [5] has been applied. A sudden change of small amplitudes due to turbulent buffeting to large amplitudes caused by fluidelastic instability characterises the stability point that depends on the adjusted damping (Figure 1b).

1.1 Fluidelastic instability

Fluidelastic instability is a very dangerous excitation mechanism. Because of the resulting large tube amplitudes this phenomenon must be avoided in any case. To represent the stability behaviour, Connors [13] introduced a diagram (cf. Figure 1b) which shows the stability boundaries as a function of the mass-damping parameter $\delta_r = \mu\delta/(\rho d^2)$ and the reduced gap velocity $V_r = u/(f_1 d)$, where μ is the mass per unit length, δ is the logarithmic decrement of damping, ρ is the fluid density, u is the velocity in the gap between the tubes and f_1 is the first natural frequency of the tube. This standard stability diagram gives a general idea of the stability behaviour of tube arrays in cross-flow which is essential for heat exchanger design. Since a reliable theoretical model is missing an experimental investigation of the stability boundaries is required.

Fluidelastic instability appears in two different mechanisms: as fluid-damping-controlled (galloping) and as fluidelastic-stiffness-controlled instability. The galloping mechanism leads generally to tube vibrations in cross-flow direction. For the tube excitation those components of the fluid force are important which are proportional to the tube velocity. In case of the fluidelastic-stiffness-controlled mechanism the instability results from coupling effects of several tubes in an array by the fluid. Here, the stability investigations requires a fully flexible bundle of tubes.

Which of these two mechanisms is dominant depends on the configuration of the tube array as well as on the fluid density. It is also possible that a combination of both effects occurs. In case of galloping instability the equilibrium position of *one single flexibly mounted tube* in an otherwise fixed array becomes unstable. This mechanism can be *isolated* using such a single tube. Moreover, for certain array configurations, where the galloping mechanism is dominant, there is a coincidence of the stability boundaries with that of a fully flexible bundle. Thus, with restrictions, the stability of a fully flexible bundle can be checked by the stability of a single flexibly mounted tube, cf. Lever & Weaver [21] and Austermann & Popp [5].

Simple analytical models to describe the dynamical behaviour of tube bundles which require no or only limited experimental data are rare. Price [29] gives a comprehensive overwiew about theoretical models for fluidelastic instability.

A theoretical model that describes the stability boundaries with respect to the galloping mechanism has been introduced by Lever & Weaver [20]. This analytical model is based on fully developed non stationary flow channels between the tubes which cause instability. Therefore, the aim of one part of the recent studies is to *disturb* these flow channels by tripwires and turbulence grids without changing the flow capacity and to investigate the influence on the stability behaviour. The observed damage in the first few rows leads also to the question of the influence of *increased turbulence* at the inlet of the bundle on the stability behaviour of the tubes.

1.2 Turbulent buffeting

Turbulent buffeting occurs in tube bundles subjected to cross-flow because the first tube rows of a bundle are working as turbulence generators. Thus, within the first few rows inside a tube array turbulence intensity increases downstream, cf. Païdoussis et al. [25], [28], or Sandifer & Bailey [33]. The resulting excitation mechanism causes small amplitudes and represents the reason for wear and fatigue in heat exchanger tube bundles. Here, for the design of heat exchangers as well as for the calculation of lifecycles the tube amplitudes are important quantities. Half-empirical formulations for the calculation of the amplitude behaviour as a function of the gap velocity are given by Blevins [10], [11], Axisa et al. [6], [7] and Oengoeren [24]. To calculate the amplitudes due to turbulent buffeting and to get informations about fluidelastic instability it is useful to determine experimentally the fluid forces acting on the tubes. There are two different possibilities to investigate these fluid forces, either they can be measured as integral values from the reaction forces, cf. e. g. Axisa [6] or they can be gained integrating the measured pressure distribution on the surface of the tube, cf. Blevins [10] and Mohr [23]. Similar investigations can also be found in Michel [22] and Gog [16].

2 Experimental apparatusses

The experimental investigations have been carried out at extended set-ups used by Andjelić & Popp [3], Austermann & Popp [5] as well as Romberg & Popp [30]. The arrays consist of aluminium tubes with an outside diameter of d=80mm and a length of l=800mm. The arrays are mounted within the test section of a wind tunnel. The fixings in the side walls of the test section allow to realize the tube patterns which are usually found in heat exchangers, cf. Table 1, where P is the distance of the tube midpoints. The expressions 'triangle' and 'square' mean the geometric unit cell of a configuration, 'rotated' and 'normal' shows the orientation to the cross-flow, cf. Table 1 and Figure 2. For most of the array

Figure 2: One of the used experimental set-ups, e. g. a rotated triangular array

configurations one single flexibly mounted tube in an otherwise fixed bundle has been investigated. This tube is equipped with a linear iso-viscoelastic mounting, cf. [4]. The tube is free to vibrate in streamwise (x) and cross-stream (y) direction and it is possible to vary row number, equilibrium position and damping of the tube using adjustable eddy current damper. As a matter of fact the stability boundaries for a fully flexible normal triangular

Table 1: Tube bundle configuration investigated here

Symbol	Configuration	P/d	One single flexible tube in a fixed array	fully flexible bundle
Δ :	rotated triangle	1.375	×	
Δ :	rotated triangle	1.25	×	
$\triangleleft$:	normal triangle	1.25	×	
$\triangleleft$:	normal triangle	1.25		×
$\Box$:	in-line square	1.25	×	
$\Diamond$:	normal square	1.25	×	

bundle are not in accordance with that of only one flexibly mounted tube. Therefore, the investigations for the normal triangular array with $P/d = 1.25$ are carried out at a fully flexible bundle, where 9 tubes have variable damping and linear iso-viscoelastic mountings. The tube motions in both apparatuses have been determined by a set of strain gauges.

To determine the exciting fluid forces the flexible tube inside the fixed array is equipped with 44 pressure sensors, cf. Figure 3. The sensors are placed directly under the surface

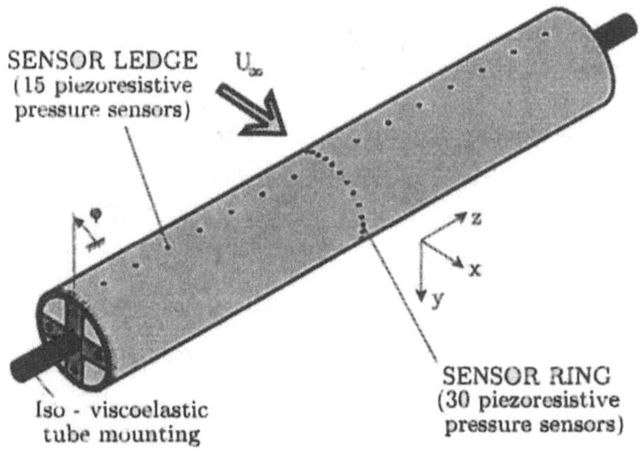

Figure 3: Sketch of the test tube

inside the test tube to avoid cavity. The acting fluid force is determined by integrating over the whole tube surface, cf. Romberg & Popp [32]. The tube motions in $x-$ and $y-$direction can be measured together with the acting fluid forces.

3 The influence of turbulence

The stability behaviour for several tube configurations visualized by means of Connors diagrams can be found in Andjelić [2] and Austermann [4] together with attempts of a theoretical prediction on the basis of the model of Lever & Weaver [20], [21]. Figure

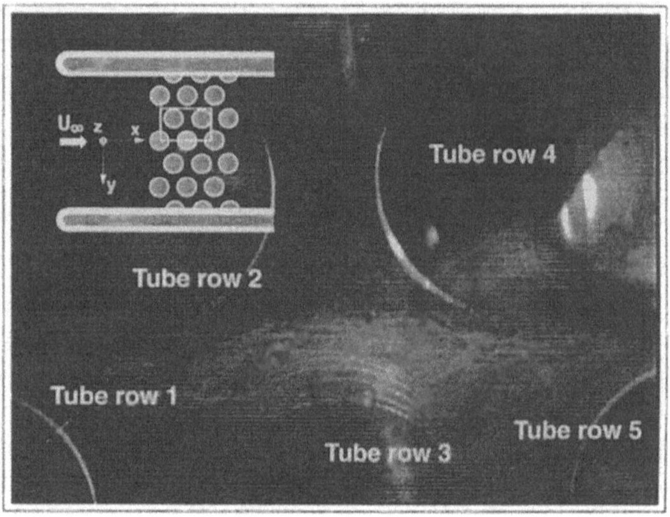

Figure 4: Oscillating pressure test tube within the visualized flow field in the third row of a rotated triangular array with $P/d = 1.375$

4 shows a photo of the excited test tube within the flow scanned out of a video film, where the flow field is visualized by a laser light sheet procedure, cf. Romberg & Popp [30]. The single flexible tube oscillates exclusively in cross-flow (lift) direction due to the fluid-damping-controlled (galloping) mechansim.

In the following the influence of a disturbed flow due to thin tripwires as well as an increased turbulence due to grids on this common behaviour is described for one configuration.

3.1 Influence of disturbed flow due to tripwires

Figure 5 shows the fully developed non stationary flow channels for a rotated triangular tube configuration assumed by Weaver & Yetisir [37]. These channels are essential for the 'Wavy wall rigid channel model' introduced by Lever & Weaver [20]. This simple linear analytical model provides the stability boundaries of a flexibly mounted tube in an otherwise fixed array in the form of

$$\delta_r = F(V_r). \qquad (1)$$

The model assumes a non stationary cross-section $A(s,t)$ of an one-dimensional flow

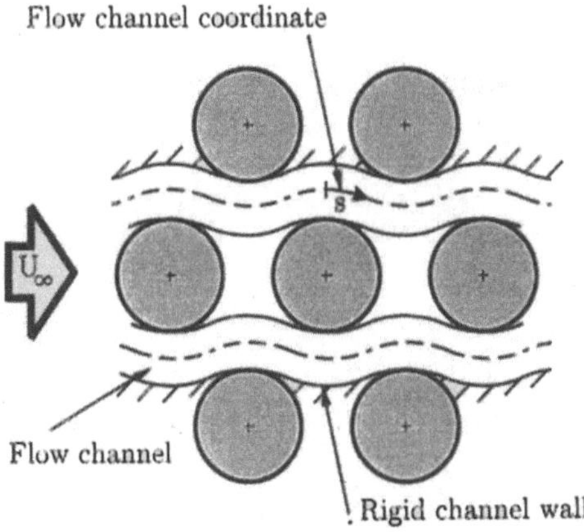

Figure 5: Flow pattern in the 'Wavy wall rigid channel model', by Weaver & Yetisir [37].

channel. The non stationary continuity equation yields the flow velocity $u(s,t)$ and the non stationary Bernoulli equation provides the pressure distribution $p(s,t)$. The integration over the pressure distribution $p(s,t)$ on the tube surface delivers the fluid force for the equation of motion of the tube. Here, the locations of flow-attachement and flow-separation are decisive parameters. If the term of the total damping disappears the stability boundary of the tube is reached. This results in eq. (1).

To influence the flow, very thin wires (1.25% of the tube diameter d) are pasted on the top and the bottom of each tube of a bundle parallel to the tube axis.

Figures 6 and 7 show the flow field in the case of an undisturbed and a disturbed flow, respectively for a rotated triangular configuration. The visualization of the undisturbed flow has been carried out for all mentioned tube array configurations at Reynolds numbers in the range of $Re = 1000$ to $Re = 100000$. Distinct flow channels have been found only for *aligned* array configurations. The results are documented in video films and photographs. Figure 6 shows the undisturbed flow pattern for a tube array with $P/d = 1.375$ scanned out of a film. In the middle of the lower half of Figure 6 the tube in the third row can be seen. The photo illustrates the free undisturbed flow channel that cause fluid-damping-controlled (galloping) instability.

Figure 7 shows the corresponding disturbed flow, where tripwires have been pasted on each tube in the bundle. On the surface of the third row one of the tripwire can be seen which influences the flow channel visibly. The flow seems to be more turbulent as well as more attached. The corresponding film shows also the stabilization of an oscillating tube due to pasted tripwires.

The first documented attempt to disturb the flow field around rigid bodies by using tripwires is due to Prandtl [27]. He put a thin wire on the surface of a ball and found a tremendous change of the flow detachement due to disturbances of the boundary layer. Further investigations with tripwires can be found in [1], [14], [17], [18], [19] and [26].

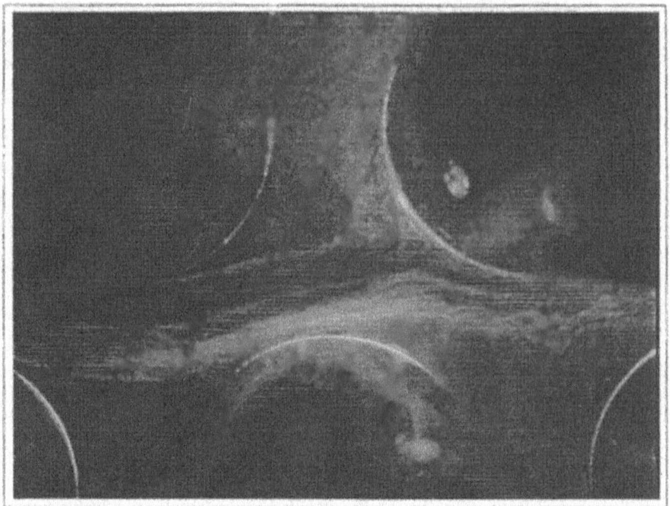

Figure 6: Visualization of the flow channel in the first four rows of a rotated triangular array with a pitch-to diameter ratio P/d =1·375 for *undisturbed* flow, $Re \sim 25000$.

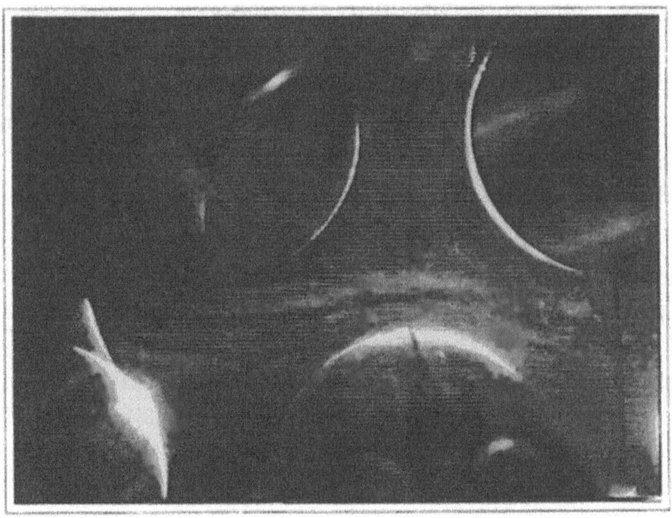

Figure 7: Visualization of the flow channel in the first four rows of a rotated triangular array with a pitch-to diameter ratio P/d =1·375 for *disturbed* flow due to tripwires, $Re \sim 25000$.

In the following, the significant influence of the disturbed flow channels on the stability behaviour is shown again for the rotated triangular array with $P/d = 1.375$. Figure 8 shows the measured reduced amplitudes for undisturbed flow as a function of the reduced gap velocity V_r and the mass-damping parameter δ_r for the same tube and ideal geometry. In the stable region the tube moves with small amplitudes in cross- and in-flow direction (turbulent buffeting). In the unstable mode the tube executes galloping motions exclu-

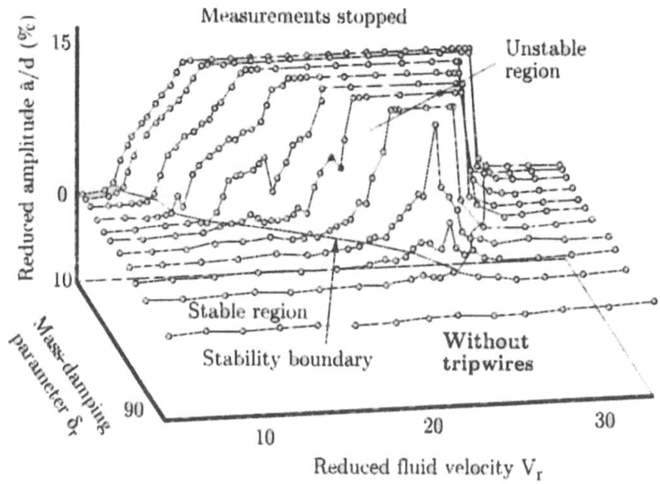

Figure 8: Measured reduced amplitudes as a function of V_r and δ_r for one single flexible mounted tube in the critical third row of an otherwise fix ed array with ideal geometry, without tripwires.

sively in cross-flow direction. To protect the experimental set-up the measurements have been stopped if the tube reaches a reduced amplitude of about 12%. After the application

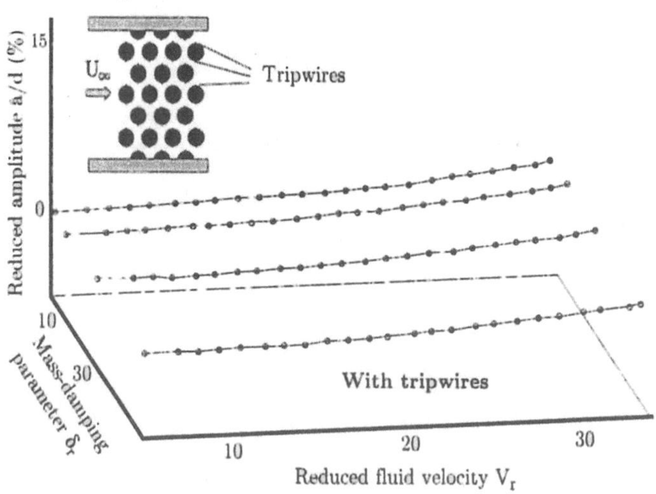

Figure 9: Measured reduced amplitudes as a function of V_r and δ_r for one single flexible mounted tube in the critical third row of an otherwise fixed array with ideal geometry, with tripwires.

of the thin Prandtl-tripwires at the same experimental set-up with the same array configuration and parameters, the system shows the amplitude behaviour presented in Figure 9. It can be seen easily that the unstable mode is fully suppressed. Small distortions of the tripwire position due to tube rotations lead to the same stabilization effect.

However, the tripwires were pasted solely on the single flexible tube, this leads to a slight shift of the stability boundary to higher reduced gap velocities only but does not change the stability behaviour. The tripwires pasted on the tubes which lie upstream to the test tube have much more influence than tripwires on the test tube. A calculation using the Lever & Weaver-model with changed geometry parameters due to tripwires does *not* show the trend of a stabilization. The investigations were carried out for several bundle configurations (also for a fully flexible bundle) as well as several shifts of the equilibrium position of the test tube. The different results are summerized as Connors diagrams in Romberg & Popp [30].

The influence on the stability behaviour of a flow disturbance *inside* a tube bundle leads to the question of the influence of increased turbulence of the upstream flow.

3.2 Influence of increased upstream turbulence

To vary the turbulence field in the test section of a wind tunnel, turbulence grids are generally used. Here, a grid is placed in front of the tube array at a distinct but variable distance X, cf. Figure 10. Figure 11 shows a photo of the experimental set-up where at the left hand side one of the grids made of wood can be seen. The grid yields a broadband

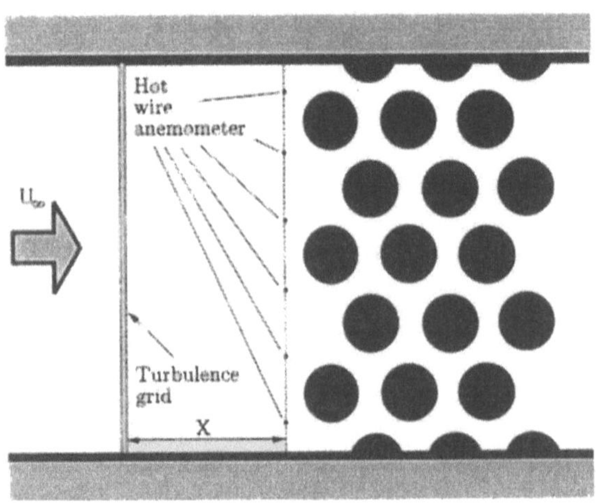

Figure 10: Experimental set-up for increasing the turbulence at the inlet of the tube bundle.

turbulence distribution. The turbulence has been measured using hot wire anemometers. In this study the turbulence is characterized by two parameters, the turbulence intensity Tu (velocity-fluctuation related to the mean velocity) and the scale length L_x (averaged length of a turbulence eddy). Details about the theoretical description of the turbulence field can be found in Romberg & Popp [31]. To describe the field in front of the bundle the hot wire was placed in 24 points vertical to the flow at the inlet of the tube array, cf. Figure 10.

Figure 11: Test array with mounted turbulence grid in front of the tube bundle.

The turbulence behaviour of the flow behind a grid has been investigated by many authors, e. g. Baines & Peterson [8], Soper [34] and Southworth & Zdravkovich [35]. In [8] it is shown for grids and screens that homogeneous flow is given at a distance of 20 bar width behind a turbulence grid. At this distance maximum turbulence intensity Tu can be observed. With increasing grid distance X the turbulence intensity Tu decreases and the scale length L_x grows. The experimental investigations in [31] show a good agreement with the theoretical consideration of the decay of turbulence by Frenkiel [15]. Depending on the used grid and the investigated bundle configuration homogeneous isotropic turbulence from $Tu = 3.5\% - 23\%$ have been generated. The $Tu-$values are largely independent from the fluid velocity U_∞ in the considered parameter region. However, the grid distance X, the mesh size s, the bar width a, the geometry of the bars and the tube bundle itself have an influence on the turbulence intensity Tu and the scale length L_x.

Figure 12 shows the corresponding three dimensional stability diagram for increased turbulence again for the configuration described above (third row, rotated triangular array, $P/d = 1.375$). Here also a significant stabilization due to increased turbulence can be seen. The summerized results concerning increased turbulence for other tube rows as well as for a fully flexible bundle can be found in Romberg & Popp [31].

To find an explanation for the effects due to tripwires and increased turbulence, in the following we take a look at the fluid forces acting on the tubes.

4 Investigation of the acting fluid forces

With the pressure test tube, cf. Figure 3, it is possible to determine the non-stationary fluid force vector $\mathbf{f}(t)$ as a force per unit length in tube midplane as well as the correlation length L_c for each angle φ at the circumference. The fluid force vector $\mathbf{f}(t)$ is calculated

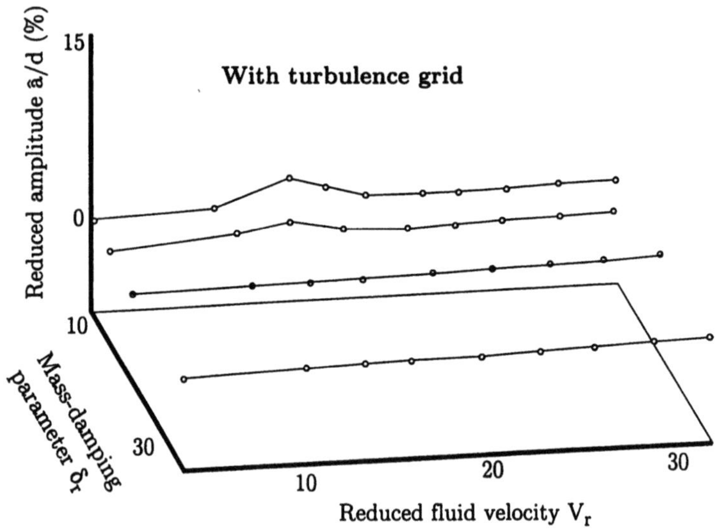

Figure 12: Measured reduced amplitudes as a function of V_r and δ_r for one single flexibly mounted tube in the third row of an otherwise fixed rotated triangular array with increased turbulence

by integrating the simultaneously measured pressure distribution $p(z = 0, \varphi, t)$ over the whole circumference. The correlation length L_c is coming from a statistical derivation, where the cross-correlation functions between the pressure signals $p(z, \varphi_i, t)$ somewhere on the tube surface and a reference signal $p(z = 0, \varphi_i, t)$ in tube mid plane are determined. Therefore, the fluid force $\mathbf{F}(t)$ acting on the whole tube can be considered as an excitation. Details of the theoretical description of the fluid forces can be found in Romberg & Popp [32].

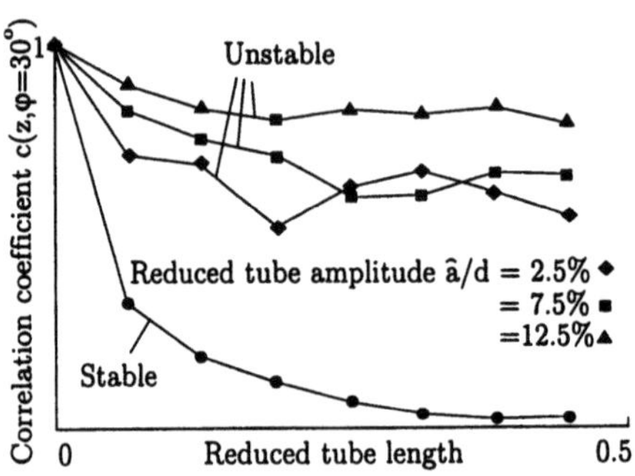

Figure 13: Local correlation coefficient c at the stability boundary for several stable limit cycles and $\varphi = 30^o$

The experimental results show that for the *stable* parameter region the spatial correlations are nearly identical for a flexibly and a rigid mounted test tube. Moreover, it has been found that these values are largely independent from the reduced velocity V_r.

Beyond the stability boundary, where fluidelastic instability due to galloping appears, suddenly the correlation lengths increase to very large values, cf. Figure 13. This holds for all investigated tube configurations where galloping can be found. Tests in still air show clearly that this is *not* due to inertia effects inside the pressure transducers or the cross-flow velocity of the tube. The dynamic pressure due to the tube velocity is 2 to 4 orders of magnitude lower than the measured pressure fluctuation inside the flow field.

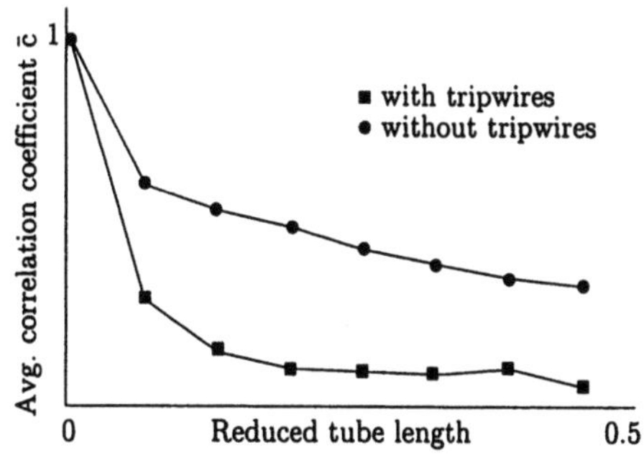

Figure 14: Averaged correlation coefficient c as a function of the reduced tube length z/l for a tube in the third row of a rotated triangular array with $P/d = 1.375$ with and without tripwires

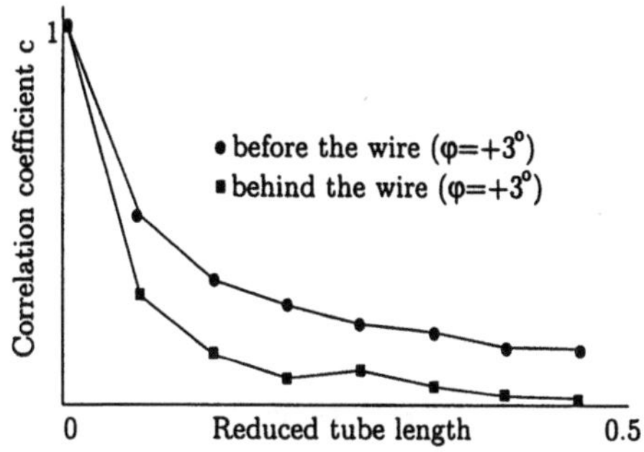

Figure 15: Correlation coefficient c as a function of the reduced tube length z/l for a tube in the third row of a rotated triangular array with $P/d = 1.375$ measured directly before and behind a pasted tripwire

If we consider the 'route to instability', it can be found that the correlation increases already if *small* stable limit cycles occur, cf. Figure 13. The fluctuation for the smallest amplitude is due to the difficulty of adjusting such a small stable limit cycle for the whole recording time. The measured fluid forces show harmonic character and lead to a classical resonance behaviour of the tube. Finally, it has been found that beyond the stability boundary, where the fluid-damping-controlled mechanism (galloping) occurs, the correlation length of the acting fluid forces is in the order of magnitude of the tube length.

A possible explanation for the stabilizing effect due to flow disturbance and increased turbulence is a reduction of the correlation length. Recent studies shows that tripwires seem to destroy the large correlation length while galloping is dominant. Figure 14 shows the correlation coefficient of the pressure averaged over the whole circumference. The investigations with turbulence grids show similar results. By now it is still not clear whether the suppression of the correlation is due to the presence of the tripwires or just a consequence of the absence of the big amplitudes because of the pasted wires. To solve this problem further experimental investigations have to be carried out. However, a hint for a possible suppression of the correlation due to the tripwires is given by Figure 15.

5 Conclusions

Experimental investigations of the stability behaviour of tube bundles subjected to cross-flow have been carried out. The experimental set-up allows to realise several bundle configurations with one single flexibly mounted tube in an otherwise fixed array as well as with a fully flexible bundle for a certain array configuration.

The presented results show clearly that a flow disturbance inside the bundle due to tripwires on the tube surfaces as well as an increased turbulence intensity at the inlet of the bundle using turbulence grids can suppress the dangerous fluid-damping controlled (galloping) instability of a flexible tube in a bundle. These effects can be of interest for designers and operators of heat exchangers.

Further investigations with respect to the acting fluid forces as well as an experimental determination of the stability behaviour due to changed flow conditions for fully flexible bundles with other configurations are necessary.

Acknowledgement

The authors gratefully acknowledge the financial support of this research work by the DFG (Deutsche Forschungsgemeinschaft), contract number Po 136/17/1+2).

References

[1] AIBA, S. 1992 Heat transfer around a circular cylinder with a single trip wire. In *Heat Transfer - Japanese Research*, Vol. 21 No. 5, pp. 466-482.

[2] ANDJELIĆ, M. 1989 Stabilitätsverhalten querangeströmter Rohrbündel mit versetzter Dreieckteilung. Ph. D. Thesis, Hannover, University, Fakultät für Maschinenwesen, Inst. f. Mechanik, in German.

[3] ANDJELIĆ, M. AND POPP, K. 1989 Stability effects in a normal triangular cylinder array. *Journal of Fluids and Structures* **3**, 165-185.

[4] AUSTERMANN, R.: 1993 Zum Schwingungsverhalten querangeströmter Rohrbündel bei Anregung vom Galloping-Typ. Ph. D. Thesis, Hannover, University, Fakultät für Maschinenwesen, Inst. f. Mechanik, in German.

[5] AUSTERMANN, R. AND POPP, K. 1995 Stability behaviour of a single flexible cylinder in rigid tube arrays of different geometry subjected to cross-flow. *Journal of Fluids and Structures* **9**, 303-322.

[6] AXISA, F.; ANTUNES, J.; VILLARD, B.; WULLSCHLEGER, M. 1988 Random excitation of heat exchanger tubes by cross-flow. *Proc. Int. Symp. on Flow-Induced Vibration and Noise*, Chicago, Vol. 2, Flow-Induced Vibration of Cylinder Arrays in Cross-Flow, pp. 23-46.

[7] AXISA, F.; VILLARD, B. 1992 Random excitation of heat exchanger tubes by two-phase cross-flow. *Proc. Int. Symp. on Flow-Induced Vibr.* Anaheim, Vol. 1, pp. 119-140.

[8] BAINES, W. D.; PETERSON, 1951 An Investigation of Flow Through Screens. In: *Transactions of the ASME 73*, S. 467-480

[9] BLEVINS, R. D.; GIBERT, R. J.; VILLARD, B. 1981 Experiments on vibration of heat exchanger tube arrays in cross flow. *Trans. 6th Int. Conf. on Struct. Mech. in Reactor Techn.*, Paris, Paper B6/9.

[10] BLEVINS, R. D. 1990 Flow-induced vibration *Van Nostrand Reinhold* New York, 2. Edition.

[11] BLEVINS, R. D. 1994 Turbulence-induced vibration of heat exchanger tubes in cross flow *ASME PVP-Vol. 273, Flow-Induced Vibration*, pp. 199 - 210.

[12] CHEN, S. S. 1988 Some issues concerning fluidelastic instability of a group of circular cylinders in crossflow. In *Proceedings of 1988 International Symposium on Flow-Induced Vibration and Noise.* Vol 3: Flow Induced Vibration and Noise in Cylinder Arrays, (ed. by M. P. Païdoussis, S. S. Chen & M. D. Bernstein), pp. 1-24.

[13] CONNORS, H. J. 1970 Fluidelastic vibration of tube arrays. *Symposium of Flow Induced Vibrations in Heat Exchangers*, New York, pp. 42-56.

[14] FAGE, A. AND WARSAP, J. 1929 The effects of turbulence and surface roughness on the drag of a circular cylinder. In: *Aeronautic Research Council R&M No. 1283.*

[15] FRENKIEL, F. N. 1948 The Decay of Isotropic Turbulence. In: *Transactions of the ASME 70*, (1948), S. 311-321

[16] GOG, W. 1982 Untersuchungen der Erregermechanismen am Einzelrohr und am querangestroemten, schwingenden Rohrbuendel mit Dreieckteilung *Ph. D. Thesis, Berlin, Techn. Univers.*, in German.

[17] HEINECKE, E. P. 1978 Fluid-elastic vibrations in heat exchangers with tubes in cross-flow. *Keswick U.K. Session 2:* Flow induced vibrations in heat exchangers.

[18] JAMES, D. AND TRUONG, Q.-S. 1972 Wind load on cylinder with spanwise protrusion. *Journal of the Engineering Mechanics Division* **1**, 72-81.

[19] KRAEMER, K. 1959 Umschlagmessungen hinter Stolperdrähten. In *Bericht 59/A/21, Aerodynamische Versuchsanstalt Goettingen,* Goettingen, Germany, (in German).

[20] LEVER, J. H. AND WEAVER, D. S. 1982 A theoretical Model for Fluidelastic Instability in Heat Exchanger Tube Bundles. *Journal of Pressure Vessel Technology* **104**, 147-158.

[21] LEVER, J. H. AND WEAVER, D. S. 1986 On the stability of heat exchanger tube bundles, Part I: Modified theoretical model. *Journal of Sound and Vibration***107**, 375-392.

[22] MICHEL, A.; HEINECKE, E.; VON DER DECKEN, C. B. 1986 Messung von instationaeren Stroemungskraeften in fluchtenden und versetzten Rohrbündeln bei starren und schwingenden Rohren. *Ph. D. Thesis, D 82*, Aachen, T.H., Berichte der Kernforschungsanlage Juelich, Nr. 2038, ISSN 0366-0885, in German.

[23] MOHR, K.H. 1981 Messungen instationaerer Druecke bei Queranstroemung von Kreiszylindern unter Beruecksichtigung fluidelastischer Effekte *Ph. D. Thesis, Juelich, Kernforschungsanlage GmbH*, IRB, Juel-1732, in German.

[24] OENGOEREN, A.; ZIADA, S. 1992 Unsteady fluid force acting on a square tube bundle in air cross-flow. *Int. Symp. on Flow-Induced Vibration and Noise*, Anaheim, Vol. 1, pp. 55-74.

[25] PAÏDOUSSIS, M. P.; PRICE, S. J.; MACDONALD, R.; MARK, B., 1987 The flow-induced vibration of a single flexible cylinder in a rotated square array of rigid cylinders with pitch to diameter Ratio of 2.12. *Journal of Fluids and Structures*, 1, pp. 359-378.

[26] PEARCEY, H., CASH, R. AND SALTER, I. 1982 Flow past circular cylinders: Simulation of full-scale flows at model scale. *NMI R 131 and OT-R-8208.*

[27] PRANDTL, L. 1914 Der Luftwiderstand von Kugeln In -*Nachr. Ges. Wiss. Goettingen*, Goettingen, Germany, pp. 177-190, (in German).

[28] PRICE, S.J.; PAÏDOUSSIS, M.P.; MACDONALD, R.; MARK, B., 1986 The flow-induced response of a single flexible cylinder in an Array of rigid cylinders: a comparison between air- and water-flow results. *Proc. Pressure Vessels and Piping Conf. and Exhibition*, Chicago,- PVP-Vol. 104, S. 107-118.

[29] PRICE S.J., 1995 A review of theoretical models for fluidesastic instability of cylinder arrays in cross-flow. *J. of Fluids and Structures*, Vol. 9, pp. 463-518.

[30] ROMBERG, O.; POPP, K., 1996 The influence of tripwires on the stability of tube bundles subjected to cross-flow. *ASME Pressure Vessels and Piping Conference*, Montreal, Canada, Vol. 328, pp. 3-10.

[31] ROMBERG, O.; POPP, K., 1996 The influence of upstream turbulence on the stability of tube bundles subjected to cross-flow. *ASME Pressure Vessels and Piping Conference*, Montreal, Canada, Vol. 328, pp. 11-18.

[32] ROMBERG, O.; POPP, K., 1997 Random excitation by fluid forces acting on a single flexible tube in bundles subjected to cross-flow. To be published: *ASME 4th International Symposium on FSI, AE & FIV + N*, Dallas, Texas, USA, (Nov 1997).

[33] SANDIFER, J.B.; BAILEY, R.T., 1984 Turbulent buffeting of tube arrays in liquid crossflow. *Proceedings Symposium on Flow-Induced Vibrations, ASME*, Winter Annual Meeting, New Orleans, Vol. 2, pp. 211-226.

[34] SOPER, B. M. H. 1982 The Effect of Grid Generated Turbulence on the Fluidelastic Instability of Tube Bundles in Cross-Flow. In: *Taborek, J.; Hewitt, G. F.; Afgan, N.: Heat Exchangers: Theory and Praxis*, S. 325-337

[35] SOUTHWORTH, P. J. & ZDRAVKOVICH, M. M. 1975 Effect of Grid Turbulence on the Fluidelastic Vibrations of In-Line Tube Banks in Cross-Flow. In: *Journal of Sound and Vibration 39* (4), (1975), S. 461-469

[36] WEAVER, D. S. AND FITZPATRICK, J.A., 1987 A review of flow induced vibrations in heat exchangers. In *Proceedings BHRA Conference on Flow-Induced Vibrations*, (ed. by R. King), Bowness-on-Windermere, UK, pp. 1-17.

[37] WEAVER, D. S. AND YETISIR, M., 1988 On an unsteady theory for fluidelastic instability of heat exchanger tube arrays. *Int. Symp. on FIV + N*, Chicago, Vol. 3, pp. 181-195.

Whirl Distribution Measurements in Front of the Impeller of a Centrifugal Compressor

Dipl.-Ing. I. Kassens und Prof. Dr.-Ing. M. Rautenberg

Institute of Turbomachinery, University of Hannover, Germany

Keywords. Centrifugal compressor, inlet guide vane, flow distribution, simple radial equilibrium, whirl energy, potential vortex, solid-body vortex

1. Introduction

Matching of compressor performance to different operating conditions is possible by means of e.g. adjustable inlet guide vanes (IGV) [1]. Past investigations of centrifugal compressors with an IGV and a diffuser clearly demonstrate the way in which the latter alter the compressor performance map [2,3]. In order to assess the flow in an axial suction pipe of a centrifugal compressor with an IGV it is necessary to investigate the flow conditions at this location. Hardly any usable measured flow data are available in the literature for the inlet region of centrifugal compressors with an IGV.

This paper deals with the experimental determination of the flow behind an IGV. By means of a second plane of measurement in front of the impeller it was also possible to determine the flow evolution and energy conversion processes between the IGV and the impeller. Different mass flow rates between the surge line and the choke margin in the centrifugal compressor performance map were also considered in order to determine the effects of the flow velocity in relation to the guide vanes.

In order to characterize the proportionate vortex components in the flow, whirl energies (kinetic energy in the peripheral direction) were computed from the flow measurements. In addition to whirl energies, whirl constants were also determined for a potential vortex and a solid-body vortex. The whirl constants are a measure of the intensity of the individual vortices.

2. Test compressor

Fig. 1 shows a meridional section of the centrifugal compressor test rig drawn to scale. The IGV has *11* vanes and the impeller inlet flow is axial.

The plane vanes are slightly profiled. The outer profile of the intake channel narrows beyond the centerline of the IGV. Both the outer and inner walls of the intake channel are specially highlighted in the diagrams depicting flow

measurements. The inner wall marks the boundary of the vaneless circular area in the middle of the IGV, which is blocked by a blast pipe secured by a holder in front of the IGV.
The total and static pressures as well as the flow angles were recorded in the measurement plane *L* behind the IGV by means of a 5-hole cylinder probe. The housing disc accommodating the probe traversing device as well as the outer profile of the intake channel may be rotated in the peripheral direction.
Owing to the restricted space available in the proximity of the impeller a 2-hole Cobra probe and a total pressure Kiel probe were used alternately in the plane of measurement *E* in front of the impeller. The air flow at the impeller inlet is almost axial. The impeller is designed with a front shroud and has *18* backswept blades. The air intake in deflected radially in the impeller and finally flows through a vaneless, parallel-sided diffuser into the scroll.

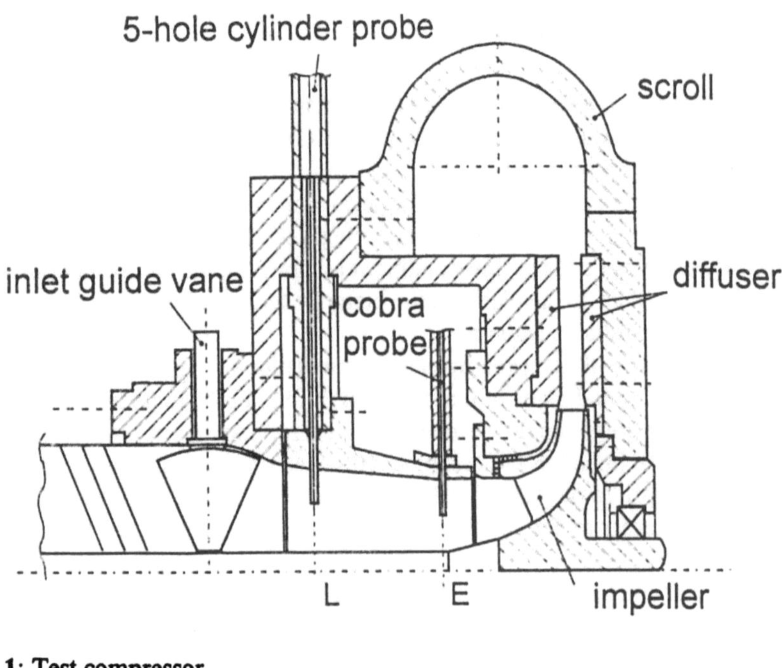

Fig. 1: Test compressor

The first measurement plane *L* behind the IGV is located at an axial distance of *1.3* times the inlet channel height and is thus at a sufficient distance from the IGV to ensure a new steady-state condition. Based on theoretical considerations and experimental investigations reported in the literature e.g. by Schröder [4] and Horlock [5], the radial compensation processes in the flow downstream of the IGV have almost completely subsided at the latter location. For this new steady-state condition it is again possible to apply the theory of simple radial

equilibrium, which presupposes constant total pressure and total temperature in the radial direction.
The reference plane of measurement *0* is located in the suction pipe in front of the blast pipe holder. The points of flow measurement in the measurement planes are shown in Fig. 2. The angular separation between the $+16.5°$ and $-16.5°$ measurement radii approximately corresponds to the IGV vane pitch. If the IGV setting is in the axial direction, corresponding to a vane angle of $0°$, one vane is located at the peripheral position $\varphi = 0°$. Measurements were made at *7* positions in the peripheral direction and at *12* positions in the radial direction.

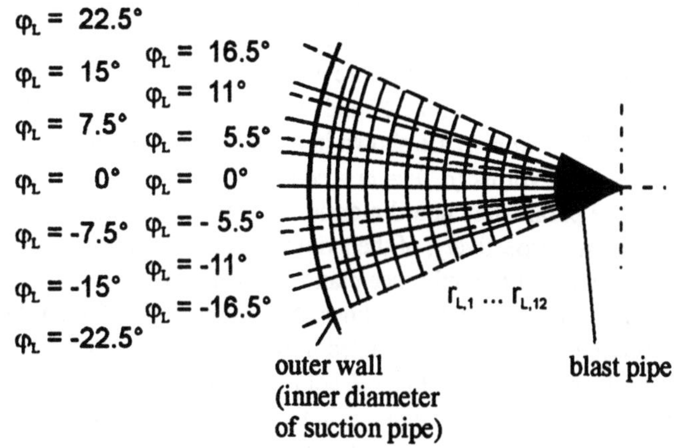

Fig. 2: Measuring points

3. Experimental results

The results of flow measurements for the example of a $60°$ IGV vane setting are first presented for the performance map reference point. The reference point is specified as the maximum isentropic efficiency at $u_2 = 225\ m/s$ and the corresponding total pressure ratio for the same mass flow rate. The mass flow rate is expressed as a percentage, whereby the mass flow rate at the maximum isentropic efficiency is defined as the *100%* mass flow rate. The radial distributions of the total and static pressure measured with a 5-hole cylinder probe in the measurement plane *L* behind the IGV as well as the computed absolute velocities for peripheral angles of $-16.5°$ to $+16.5°$ are plotted in Fig. 3. The measurement radii are referred to the inner radius r_0 of the suction pipe. The outer wall of the flow channel in the measurement plane *L* behind the IGV is located at a radius ratio r_L/r_0 of *0.8*, and the inner wall at a radius ratio of *0.17*. The inner wall signifies that the vaneless circular area in the middle of the IGV is blocked by a blast pipe. The maximum radius ratio of *0.8* in this plane of measurement illustrates the convergence of the flow-cross section behind the IGV.
A pronounced change in the total and static pressure in the radial direction is clearly evident in the upper part of Fig. 3. Especially noticeable is the large

decrease in the total pressure from about the middle of the intake channel towards the inner wall. At the last probe position close to the inner wall the total pressure behind the IGV is only *96%* of the total pressure in the suction pipe reference plane in front of the IGV. The static pressure drops uniformly from the outer wall towards the inner wall. The decrease is more pronounced from about the middle of the flow channel towards the inner wall.
The measured flow angles are plotted in the lower part of Fig. 3. Behind the IGV the flow overspins in the middle of the intake channel to *65°*, whereas the flow angle falls to about *55°* towards the channel walls. Overspinning of the flow angle is due to convergence of the cross-section. This phenomenon is known from the literature and has been reported e.g. by Iserland [6].
The whirl constants at *100%* mass flow are plotted in Fig. 4 for the four different IGV vane angles of $\alpha_{IGV} = 60°, 30°, 0°$ and $-30°$. The whirl constants for the potential vortex ($c_u * r = konst.$) are shown in the upper part of the figure while those for the solid-body vortex ($c_u/r = \omega = const.$) are shown in the lower part. In order to make these parameters non-dimensional, they were divided by the product and quotient, respectively, of the axial velocity component $c_{z,0}$ and the corresponding radius r_0 in the reference plane *0*. The inner and outer walls of the intake channel are indicated by dashed lines. At a radius ratio of *0.73* the outer wall in the measurement plane *E* is situated in front of the impeller. The solid symbols refer to the measurement plane *L* behind the IGV while the open symbols refer to the measurement plane *E* in front of the impeller.
At IGV vane angles of *30°*, *0°* and *-30°* the shapes of the curves in both measurement planes are very similar and the values for each vortex are approximately the same. This is noticeable in the whirl constants for the potential vortex as well as the solid-body vortex. The fact that the curves of the whirl constants for IGV vane angles of *-30°* and *30°* are virtually mirror images of each other clearly indicates the symmetry of the IGV vanes.
The values behind the IGV and in front of the impeller are first seen to deviate from each other at the larger IGV vane angle of *60°*. Large total pressure losses behind the IGV and in front of the impeller are then observed in the inner half of the intake channel. For all four IGV vane angles a potential vortex forms in the outer half of the intake channel. In contrast, a solid-body vortex only develops at a *60°* IGV vane angle in the inner half of the intake channel in the measurement plane *E* in front of the impeller. This combination of a potential vortex in the outer region and a solid-body vortex in the inner region of the intake channel only occurs at an IGV vane angle of *60°* in the measurement plane *E* in front of the impeller for each of the mass flows investigated.
The whirl energies for the *60°* IGV vane angle at different mass flows are plotted against the radius ratio in Fig. 5. The whirl energy represents the kinetic energy of the flow in the peripheral direction ($whirl\ energy = \frac{1}{2}\, m\, c_u^2$).
The values in the measurement plane *L* behind the IGV are plotted in the upper part of Fig. 5. The whirl energy is summed from the outer region towards the inner region and vice versa. The shape of the curves is uniform; at all mass flow

rates they rise continuously from the inner to the outer region and vice versa. A notable feature is that the radial position of the point of intersection of the curves is hardly influenced by the different mass flows.
The corresponding whirl energies in the measurement plane *E* in front of the impeller are plotted in the lower part of Fig. 5. Immediately noticeable is the irregular shape of the curves. This is due to the fact that it was not possible to measure the static pressure in the intake channel in front of the impeller for constructional reasons. The latter was determined approximately with the aid of the measured wall pressures. Despite the rather erratic behavior of the whirl energy, the overall tendency is nevertheless clear. A notable feature is that the curves from the outer to the inner region rise more sharply than those in the opposite direction. This indicates that the whirl energy is shifted towards the outer region. In overall terms the whirl energy is seen to decrease by about one-third.

4. Concluding remarks

The flow fields in the axial suction pipe of a centrifugal compressor were determined experimentally in 2 planes of measurement behind an IGV. Four different IGV vane angles were investigated. The tests were also carried out for different mass flow rates in order to determine the influence of flow velocity.
The first plane of measurement was located at a distance of *1.3* times the intake channel height from the IGV. In accordance with theoretical considerations and experimental results this is the point at which radial compensation processes downstream of the IGV have more or less subsided. The theory of the simple radial equilibrium could be applied in radial sections, when the curve of the total pressure was nearly constant. The total temperature was held constant in the radial direction. In both planes of measurement the total pressure decreases sharply towards the inner wall in the inner half of the intake channel. The measured total pressure losses of up to *10%* along the inner wall behind the IGV are quite significant for the *60°* IGV vane angle of particular interest. Several typical whirl distributions such as e.g. the potential vortex were computed from sample measurements in radial sections without such extreme total pressure losses. Unsatisfactory agreement was nevertheless obtained between the measured axial velocity distribution and the axial velocity distribution theoretically required for simple radial equilibrium ($c_z = const.$ for the potential vortex).
As the flow channel converges beyond the axis of rotation of the IGV vanes the flow angles in the plane of measurement *L* behind the IGV are larger than the vane angle settings in some cases.
On the basis of the whirl constants it was concluded that a potential vortex develops in the outer half of the intake channel for all IGV vane angles in both planes of measurement. An additional solid-body vortex develops in the inner half of the intake channel in the measurement plane E in front of the impeller only for an IGV vane angle of *60°*.

The present work should be interpreted as an initial attempt to experimentally determine the characteristics of the axial inlet flow to the impeller of a centrifugal compressor behind an IGV. Contrary to initial conjectures of a dependency between flow effects and the velocity of the flow itself, it was not possible to identify such a relationship. It appears more likely that instationary effects originating from the impeller of the centrifugal compressor influence the flow field in the upstream direction.

References

[1] Balje, O.E.: „Turbomachines: A Guide to Design, Selection, and Theory", A Wiley-Interscience Publication, John Wiley & Sons, New York - Chichester - Brisbane - Toronto, 1981

[2] Wallace, F.J., A. Whitfield and R.C. Atkey: „Experimental and Theoretical Performance of a Radial Flow Turbocharger Compressor with Inlet Prewhirl", Proceedings of the Institute of Mechanical Engineers, Vol. 189, No. 43, pp. 177-186, 1975

[3] Rodgers, C.: „Centrifugal Compressor Inlet Guide Vanes for Increased Surge Margin", ASME-Paper No. 90-GT-158, 1990

[4] Schröder, H.-J.: „Die Wirbelscheibenmethode in Anwendung auf die Auslegung axialer Turbomaschinen", Fortschr.-Ber. VDI-Z., Reihe 7, Nr. 28, VDI-Verlag, 1972

[5] Horlock, J.H.: „Axial Flow Compressors. Fluid mechanics and Thermodynamics", London: Butterwoth & Co, 1958

[6] Iserland, K.: „Untersuchungen über die Umlenkung eines freien Luftstrahls mit Hilfe von Drall", Promotionsarbeit, Eidgenössische Technische Hochschule Zürich, 1958

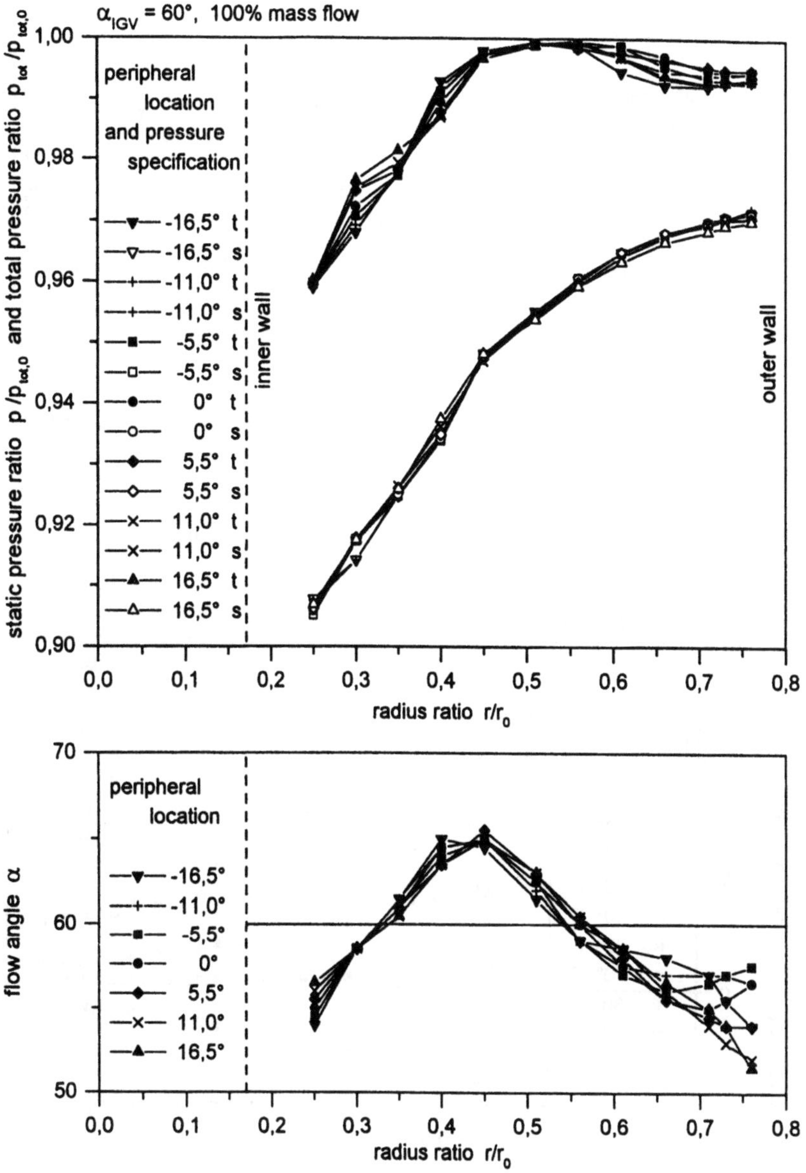

Fig.3: Total and static pressure and flow angle behind inlet guide vanes for 60° IGV-angle and 100% mass flow

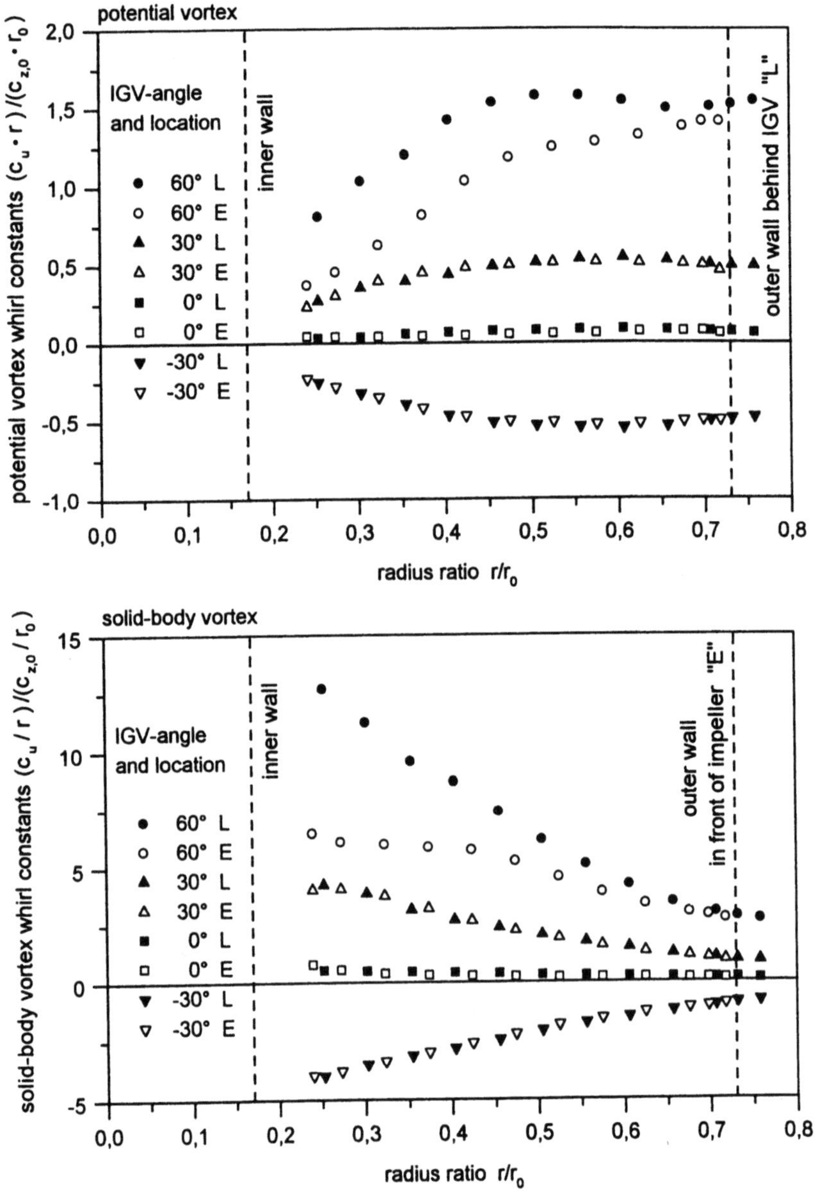

Fig.4: Whirl-constants for the potential and solid-body vortex behind the inlet guide vanes and in front of impeller for different IGV-angles

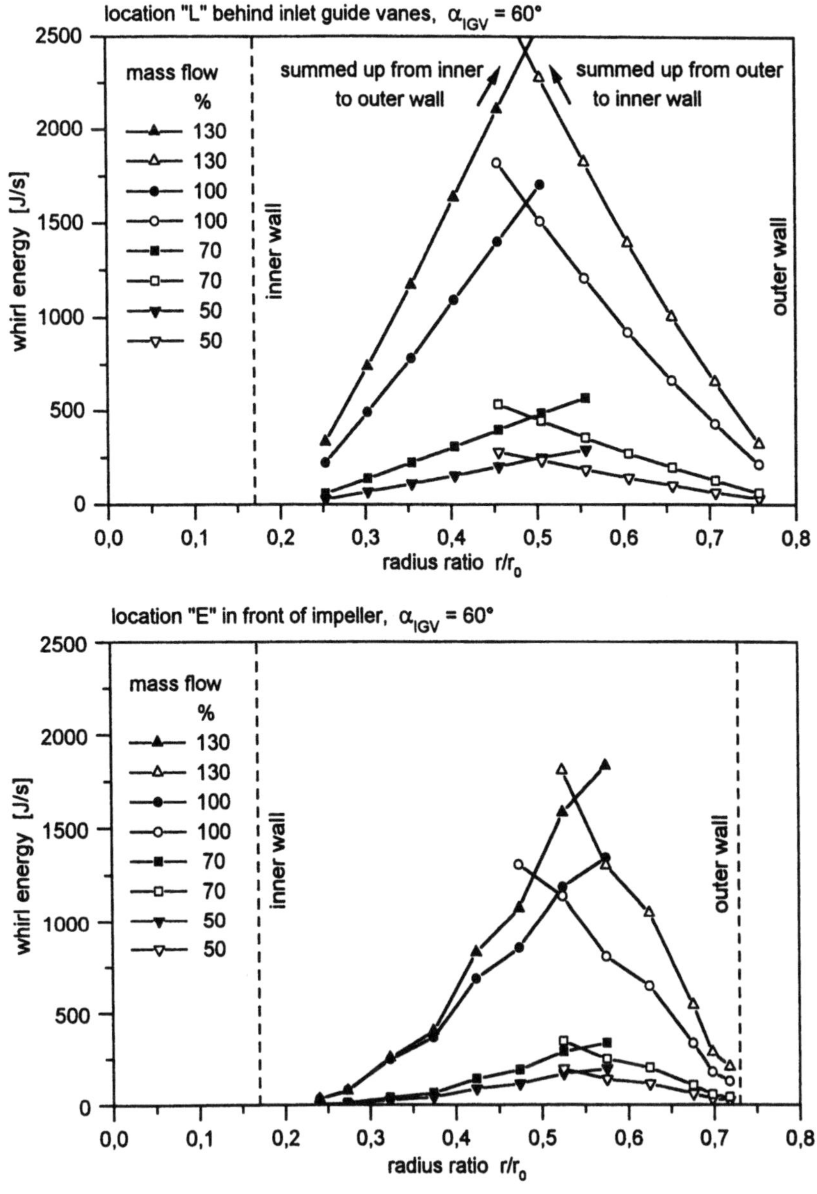

Fig.5: Whirl energies behind inlet guide vanes and in front of impeller for 60° IGV-angle and different mass flows

Velocity and Turbulence Measurements in the Flow Downstream of Flow Conditioners

Wolfgang Merzkirch, Kathrin Kalkühler
Lehrstuhl für Strömungslehre
Universität Essen
D-45117 Essen

Abstract

Hot-wire anemometry is used for measuring the flow field downstream of 3 different types of flow conditioners that are exposed to pipe flow disturbed by a 90° out-of-plane double bend. This disturbance causes asymmetry of the velocity profile and swirl.

While the swirl is found to be removed completely, the measured profiles of axial mean velocity and its turbulent fluctuactions re-develop in axial direction according to different laws. The conditioners show different efficiencies in re-developing the flow. It is apparent that the most important processes determining the efficiency of the conditioner take place in its near field which ends at a distance of about 3 or 4 pipe diameters downstream.

1. Introducation

The majority of existing types of flow meters must be calibrated. This is done in fully-developed pipe flow, and it requires that the same state of flow is established at the position where a flow meter is used for practical measurements. The existence of fully-developed flow is guaranteed after the fluid has passed through a straight pipe of certain length as defined in national and international standards (e.g. DIN, AGA, ISO). It is often claimed that the needed straight pipe lengths cannot be tolerated for practical reasons. A technical means for such a situation can be a flow conditioner whose purpose is to accelerate the re-development of the pipe flow after it has been disturbed by installations such as bends, valves, or constrictions.

A number of flow conditioners are described in the standards, but it appears that their fluid mechanical performance is not fully understood since, at least in a few cases, the standards come to the absurd result that the required length of straight pipe with conditioner is more than without conditioning. On the other hand, a few conditioners, that were designed in recent years and reported to perform more efficiently than existing devices, are not (yet) included in the standards. In this situation the present studies aim at clarifying some of the fluid mechanical problems associated with flow conditioning and making available data which may be used for comparing the efficiency of three different conditioners: the „traditional“ tube bundle (see, e.g. DIN 1952), and two perforated plates whose design has been described by Akashi et al. (1978) and Laws (1990), respectively. By means of hot-wire anemometry we have probed the flow field downstream of the three conditioners that were exposed to the flow disturbed by a 90° out-of-plane double-bend. In many investigations on installation effects, this double-bend is taken as a standard disturbance which is known to produce an asymmetry of the velocity profile and swirl (Fiedler 1997). The experimental results show the development of the velocity profile and the turbulence field in axial direction. These experiments are complimentary to measurements performed in the same facility by means of particle image velocimetry (Schlüter and Merzkirch 1997).

Flow conditioners are similar to grids and screens used in wind tunnels, and turbulence grids which serve for producing a particular type of flow turbulence. Therefore, the results obtained in our experiments will be compared with the reported performance of these devices.

2. Experiment

The experiments were performed in a pipeline with an internal diameter of D=100 mm. The internal surface of the steel pipes has a mean roughness of less than 0.5 μm and is therefore hydraulically smooth. The working fluid is air which is sucked into the pipe by means of a fan at the downstream end. It allows a maximum average pipe velocity (volume flow rate/cross-sectional area) of approximately 45 m/s, which corresponds to a pipe Reynolds number $Re_D = 3 \cdot 10^5$. The volume flow rate is determined with a specially designed measuring device, located on the downstream side of the pipeline, with an accuracy of ± 0,3%.

The air enters the pipe through a bellmouth, then flows through a straight pipe of 25D length which is followed by the „standard“ disturbance, the 90° out-of-plane double-bend (Fig. 1). 2D downstream of the bend is the position of the conditioner. The location of the test section where the hot-wire measurements are taken can be at distances between 1.5 D and 101.5 D in the straight pipe which follows the conditioner. The distance between the test section and the fan is 30 D or more.

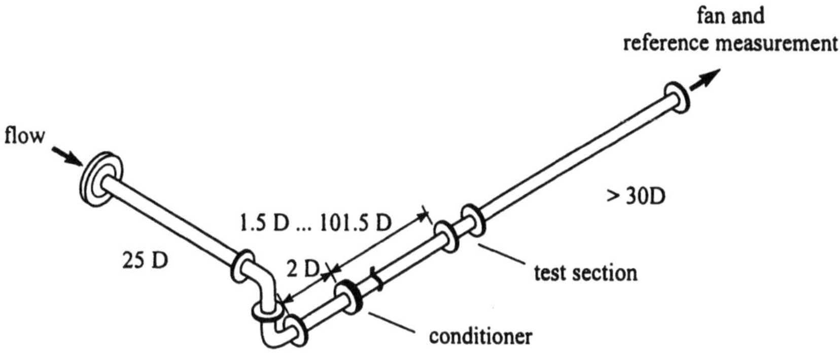

Figure 1: Pipe line test facility

The geometry and dimensions of the three conditioners investigated are shown in Fig. 2. The two perforated plates differ by the size and arrangement of the circular holes. Their values for the ratio of open area to total cross-sectional area are 0.5915 (Akashi et al.) and 0.5155 (Laws). The pressure loss coefficients, defined as $\xi = \Delta p / \frac{1}{2} \rho u^2{}_m$, are reported to be 5 (tube bundle), 1.7 (Akashi et al.), and 1.8 (Laws), which indicate the energetic advantage of the perforated plates in comparison to the tube bundle.

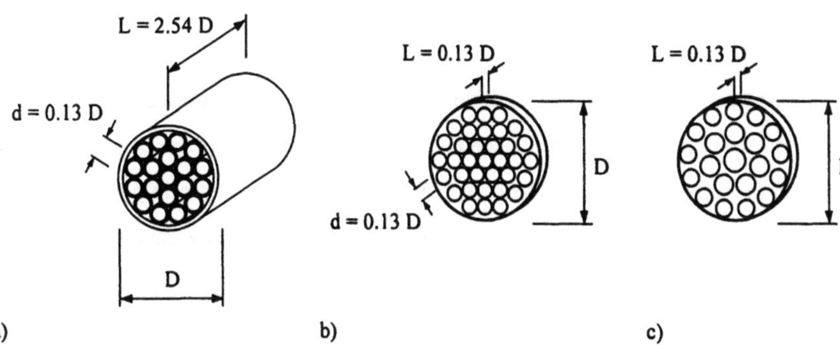

Figure 2: Geometry of conditioners: a) tube bundle, b) Akashi et al. (1978), c) Laws (1990)

For the velocity and turbulence measurements we used a DANTEC constant-temprature-anemometer with a crossed-wire probe. Calibration and measurements are controlled by a PC and the software „Streamline". In the calibration unit the probe is exposed to 13 different values of the flow angle between -30° and + 30°, and to 15 different values of the flow velocity ranging from 0.5 m/s to 55 m/s. The results of the calibration are conserved in a look-up table which is later used, with respective interpolations, for the evaluation of the measurements.

At each measuring position along the pipe axis the probe was traversed along radial paths separated by constant angular intervals, and measurements were taken at steps of 2 mm along each path, with the probe at three different orientations (0°, 45°, 90°) relative to the probe axis. Thus, the three carthesian velocity components could be determined, and they were then converted into the axial, radial, and azimuthal components. Two orientations of the prove (0°, 90°) would have been sufficient for this purpose. The third orientation at 45° served for increasing the accuracy.

The anemometer system and the applied software allowed us to determine mean (time-averaged) values of the velocity, RMS values of the turbulent fluctuations, and Reynolds stresses. Since the interest in the research programme is in the high Reynolds number range, we restricted the experiment to Reynolds numbers of $1 \cdot 10^5$, $2 \cdot 10^5$, and $3 \cdot 10^5$. Since the results for quantities presented in this article do not exhibit a noticeable Reynolds number influence, we shall show only data measured for $Re_D = 1 \cdot 10^5$.

3. Results and discussion

In order to visualize the amount of disturbance caused by the 90° out-of-plane double bend we first present data characterizing the unconditioned flow downstream of the bend. Fig. 3 shows profiles of the time-averaged axial velocity component at four downstream positions L/D and for the Reynolds number $Re_D = 1 \cdot 10^5$. The measurements were taken by traversing the hotwire probe along 4 radial paths separated by 45°. The time-averaged velocity $\bar{u}$ is non-dimensionalized by the average pipe velocity u_m. In the beginning (L/D = 1.5) the profile is highly non-symmetric. It appears to be close to fully-developed for L/D = 51.5. But this result for approaching the fully developed state only refers to the axial component. The apparent shift of the velocity maximum with varying L/D indicates the existence of swirl in this flow.

The production of swirl by the investigated bend is documented in Fig. 4, where the distribution of the tangential (azimuthal) velocity component u_t is shown in form of two vector plots for the cross-section at L/D = 3.5 and L/D = 11.5 downstream of the bend.

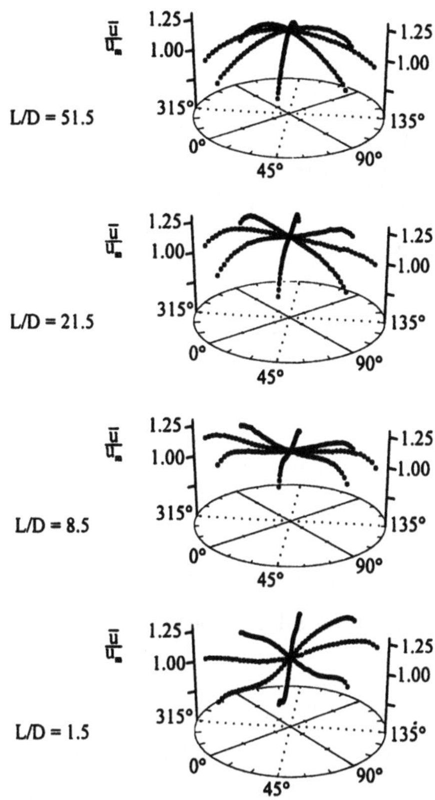

Figure 3: Time averaged axial velocity profiles at four different axial positions downstream of the 90° out-of-plane double bend.

Measurements could not be performed closer to the wall than shown due to the size and the necessary orientation of the probe for determining u_t. The maximum value of u_t/u_m observed in these cases is about 0.3. In the center of the cross-section the experimental data scatter, probably because the values of u_t and those of the turbulent fluctuations are here of the same order of magnitude. A decay of the swirl between the two measuring positions is not observed.

RMS values of the turbulent fluctuations of the axial velocity component are shown in Fig. 5. The values of u′ are made non-dimensional with the theoretical wall friction velocity u_τ for fully developed flow. Data are presented for a vertical and a horizontal measuring plane, and

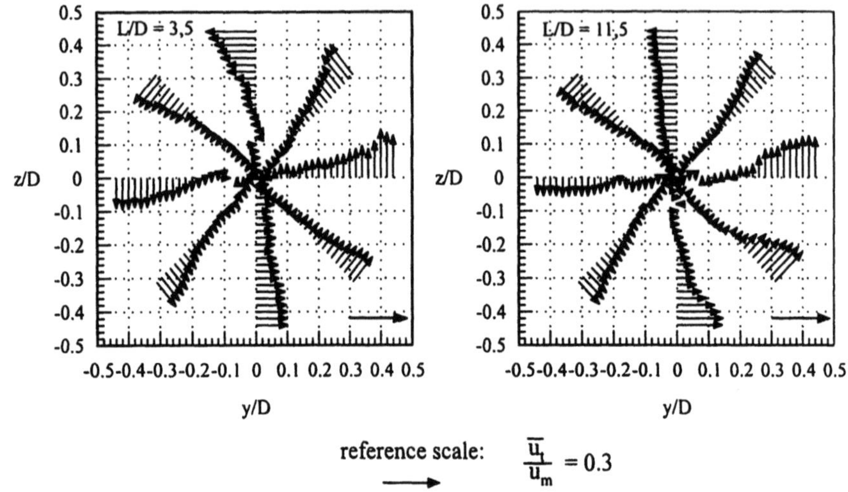

Figure 4: Vector plots of the distribution of the tangential (azimuthal) velocity component u_t in two different cross-sections downstream of the 90° out-of-plane double bend.

for 4 values of the axial position L/D. As a reference the RMS values of u'/u_τ for fully developed flow as reported by Laufer (1954) and Lawn (1971) are also included in Fig. 5. These reference data could be matched well with our respective hot-wire measuremens (see the open symbols in the lower part of Fig. 5), from which we conclude that our procedure of determining the RMS values is reliable also in disturbed flow. At L/D = 1.5 downstream of the bend the turbulence level is very high. Further downstream this level decreases, but the RMS-profiles remain asymmetric for a longer distance. Symmetry is reached at L/D= 51.5, however differences between our data and the reference values are still visible which indicates that a fully developed flow state is not yet present at this position, as one could conclude if one only regards the mean (time-averaged) axial velocity presented in Fig. 3.

The influence of the conditioners on the re-development of the axial mean (time-averaged) velocity is shown in Fig. 6. The experimental data refer to the horizontal measuring plane whose orientation with respect to the geometry of the conditioners is indicated in the upper part of the figure. Velocity profiles were determined at four axial positions L/D downstream

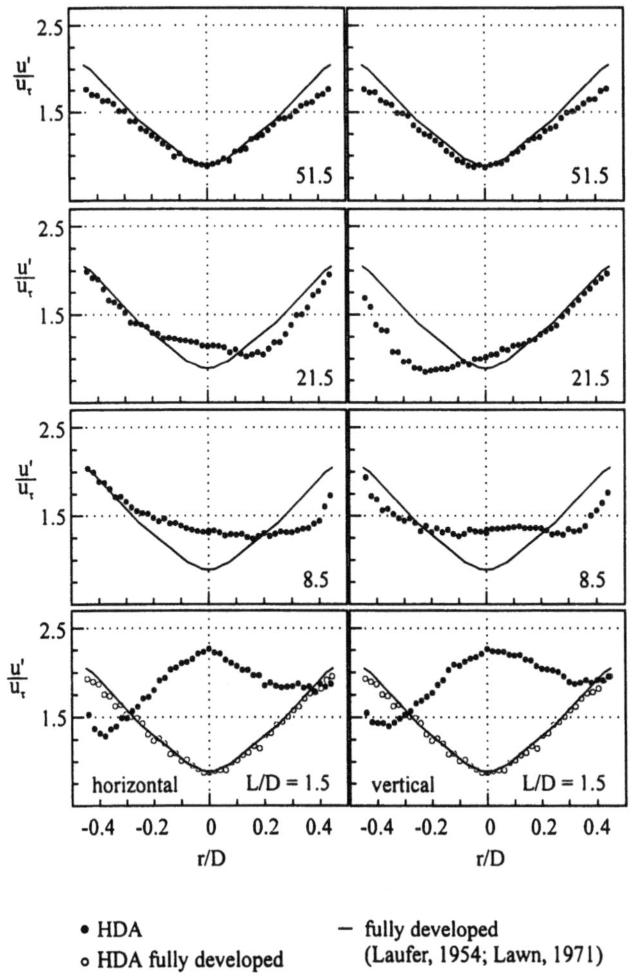

Figure 5: RMS values of axial fluctuations as measured in a horizontal and a vertical plane and at four different positions L/D downstream of the 90° out-of-plane double bend.

of the conditioners. Each of the 12 sections of Fig. 6 includes, for the purpose of comparison, the theoretical profile for fully developed pipe flow according to Gersten and Herwig (1992), and the lower three sections (L/D = 1.5) also include the profile (open symbols) that would exist at this position in the absence of the conditioners. The influence of the air jets exhausting from the openings (or tubes, respectively) of the conditioners is clearly visible in the profiles at L/D = 1.5, particularly for the tube bundle. This influence disappears almost completely at L/D = 3.5. The regime $0 \leq L/D \leq 3.5$ (or 4) apparently is dominated by strong mixing effects

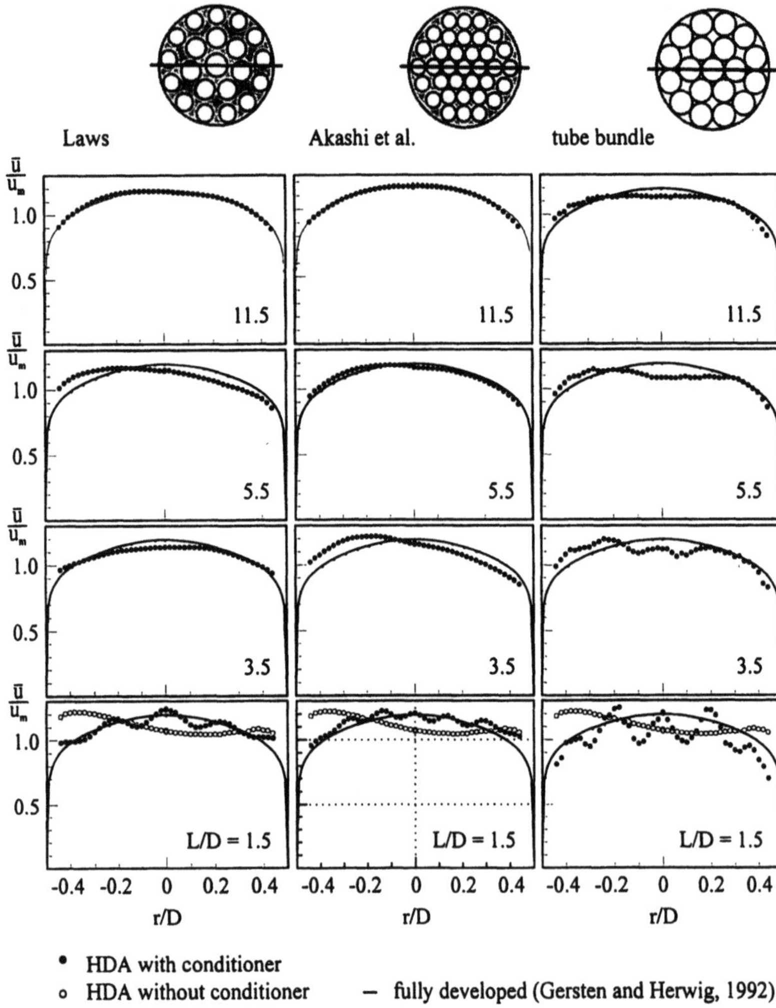

Figure 6: Time averaged axial velocity profiles downstream of the three conditioners.

caused by the interaction of the individual jets. The further re-development of the flow downstream of this position is a much slower process. At L/D = 11.5 the best result with respect to fitting the fully developed velocity profile is obtained for the Laws conditioner, whereas the tube bundle profile still exhibits considerable deviations from the fully developed state. The detailed analysis by Kalkühler (1997) shows that this favorable performance of the Laws conditioner is verified for all orientations of the measuring plane investigated.

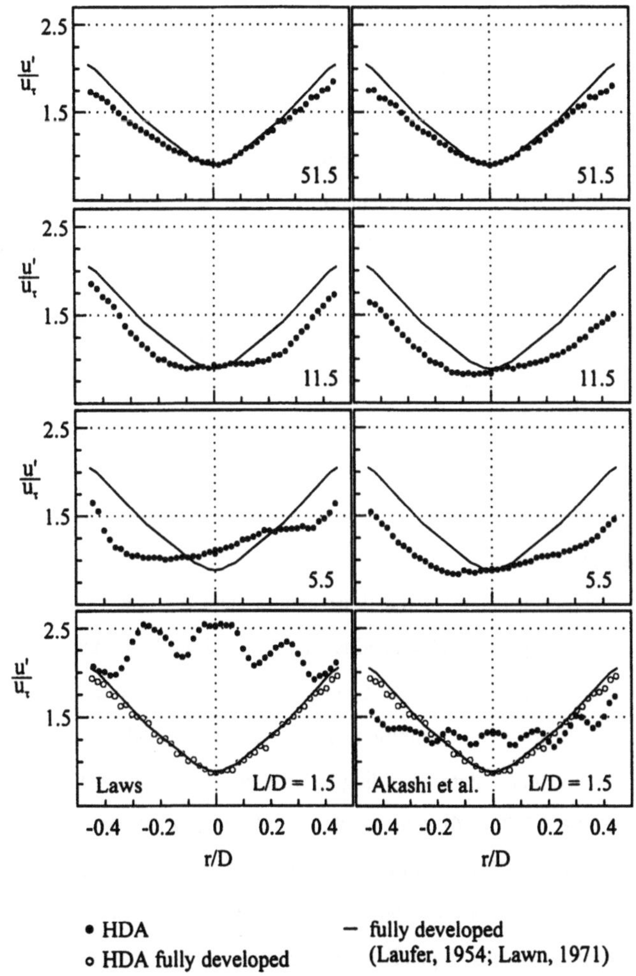

Figure 7: RMS values of the axial fluctuations downstream of the two perforated plates (Laws 1990;Akashi et al. 1978)

Downstream for the three conditioners we found the azimuthal velocity component to be zero within the limits of the measurement accuracy; i.e., the swirl caused by the 90° out-of-plane double bend (cf. Fig. 4) is removed completely.

RMS values of the non-dimensional axial velocity fluctuation, u'/u_τ, are presented in Fig. 7 for the two perforated plates. As in Fig. 4, the figure includes the data for fully developed turbulent flow for comparison. At L/D = 1.5 the results for the two conditioners differ considerably regarding the turbulence level. This must be attributed to the difference in size of the openings that are larger in the case of the Laws conditioner as compared to the conditioner of Akashi et al., or: in terms of hole diametr to pipe diameter the results at L/D = 1.5 refer to two different situations. As in the case of the mean values (Fig. 6), the influence of the individual jets decreases rapidly in downstream direction and the center of the RMS profile becomes relatively flat, whereas the influence of the pipe wall increases as it is apparent at L/D = 5.5. Most interestingly, at the far downstream position L/D = 51.5, the RMS values have not yet reached the distribution known for fully developed flow (Laufer 1954/ Lawn 1971), but the results obtained here are almost the same as those for the flow downstream of the bend without conditioner (Fig. 5). One could conclude that the disturbance (bend) imposes onto the flow a characteristic pattern of turbulence which is not removed by the conditioners.

4. Summary and conclusions

Our analysis of the performance of the three conditioners that were exposed to the flow disturbed by a 90° out-of-plane double bend is mainly based on the measurement of the axial mean and fluctuating velocity and the mean azimuthal velocity component. While the latter is virtually zero downstream of the conditioners, the axial mean and fluctuating components approach the fully developed state according to different laws. This state is reached by the mean values for L/D ≈ 10 (Fig. 6), whereas the turbulence characteristics arrive at this state only for L/D > 50 (Fig. 7). This result concerning the conditioning of the turbulence of the flow should be of interest for a possible numerical modelling of this flow, as well for flow metering by methods that are dependent on the state of the turbulence, e.g. ultrasonic correlation methods (Rettich and Poppen 1997).

From the turbulence pattern at L/D = 51.5 (Fig. 7) it follows that the influence of the conditioner is lost completely, but a comparison with Fig. 5 shows that the turbulence pattern has returned to the characteristic state caused by the disturbing bend. We may conclude that the conditioner superimposes a turbulence pattern to the turbulent field to which it is exposed by the flow without removing the incoming turbulence field.

The results presented in Figs. 6 and 7 indicate that the most violent flow processes downstream of the conditioners occur in the „near field" that terminates approximately at L/D = 4. The near field is dominated by the interaction of the jets issuing from the openings of the conditioner, and this may result in a highly unsteady flow. Similar patterns have been observed in the flow downstream of wind tunnel screens (Böttcher and Wedemeyer 1989). Bradshaw (1965) reports that unsteady flow downstream of such screens occurs if the ratio of open area (orifices) to total cross-sectional area of a screen is lower than 0.6. This situation is given for the two orifice plates investigated here; the values of these ratios are 0.59 (Akashi et al. 1978) and 0.51 (Laws 1990), respectively. It appears that optimizing the design of a conditioner is to maximize the mixing in the near-field as caused by the interacting unsteady jets, and simultaneously minimizing the pressure loss.

The conditioner imposes onto the flow its own structural characteristics. The downstream end of the near field ($L/D \approx 4$) can be taken as the position where these characteristic structures have disappeared due to dissipation. If we attribute to the conditioner an effective mean orifice diameter, then the end of the near field is at about twenty of such diameters downstream. This is in agreement with the flow through turbulence grids that are used to produce quasi-isotropic turbulence (see, e.g., Van Atta and Yeh 1972). It is assumed that the quasi isotropic turbulent flow is reached at approximately twenty mesh sizes downstream of the grid, i.e. flow structures caused by the grid geometry have decayed up to this position.

From the results of Fig. 6 we may conclude that the minimum length of straight pipe required between the conditioner and a flowmeter is 10 D. Even this value is considered too big in many practical situations. An alternative for a further reduction of straight pipe length is then to position the meter at the end of the near field and to calibrate it as a system together with the conditioner.

Acknowledgement
The authors gratefully acknowledge financial support of this research by Deutsche Forschungsgemeinschaft (DFG Az. Me 484/29).

References

Akashi, K.; Watanabe, H.; Koga, K. 1978: Flow rate measurement in pipe line with many bends. Mitsubishi Heavy Ind. 15 (2), 87-96.

Böttcher, J.; Wedemeyer, E. 1989: The flow downstream of screens and its influence on the flow in the stagnation region of cylindrical bodies. J. Fluid Mech. 204, 501-522.

Bradshaw, P. 1965: The effect of wind-tunnel screens on nominally two-dimensional boundary layers. J. Fluid Mech. 22, 679-687.

Fiedler, H.E. 1997: A note on secondary flow in bends and bend combinations. Exp. Fluids 23, 262-264.

Gersten, K.; Herwig, H. 1992: Strömungsmechanik. Vieweg-Verlag, Wiesbaden.

Kalkühler, K.: Dissertation, Universität Essen.

Laufer, J. 1954: The structure of turbulence in fully developed pipe flow. NACA Rep. 1174.

Lawn, C. 1971: The determination of the rate of dissipation in turbulent pipe flow. J. Fluid Mech, 48, 477-505.

Laws, E.M. 1990: Flow conditioning - a new development. Flow Meas. Instrum. 1, 165-170.

Rettich, T.; Poppen, G. 1997: Ultrasonic flow measurement in turbulent pipe flow using a correlation technique. Proceed. FLUCOME '97, 167-170.

Schlüter, T.; Merzkirch, W. 1996: PIV measurements of the time-averaged flow velocity downstream of flow conditioners in a pipeline. Flow Meas. Instrum. 7, 173-179.

Van Atta, C.W.; Yeh, T.T. 1972: Spectral transfer of scalar and velocity fields in heated grid turbulence. J. Fluid Mech. 58, 233-261.

Experimental Research on Axially Excited Rotating Liquid-Liquid Systems

J. Siekmann and D. Raake

Chair of Mechanics; University-GH Essen, FB 12 - Mechanik, 45117 Essen, Germany

The natural modes and natural frequencies of an incompressible and immiscible fluid system (petrolether 40/60 and water), enclosed in an axially excited and rapidly rotating cylindrical tank, are studied experimentally in dependence on different and variable parameters: the tank diameter, the geometric ratio (tank height/tank diameter), the filling ratio, the rotational tank frequency and the excitation frequency. The undisturbed cylindrical surface, being parallel to the direction of gravity, is generated by a high rotational frequency. Due to an axial excitation, a disturbance of the interface occurs. The resulting natural modes are documented photographically. In order to analyze the experimental data, dimensionless numbers are introduced, namely the Bond number, the ratio of the tank height to a mean radius of the phase boundary between the fluids, and the frequency ratio. The experimental procedure is described and the results are discussed critically.

1. Introduction

Rotating fluid systems are important in many fields of technology, and thus a study of the occurring phenomena is of considerable theoretical and practical interest. The range of the typical frequencies reaches from $1 d^{-1}$ (rotating of the earth) to $10^3 s^{-1}$ (centrifuges). Characteristic lengths of rotating systems are usually the radii. They are within a range of 10^6 m to 10^{-3} m.

---

* Based on the written and expanded version of a poster presentation at the *Ninth European Symposium* GRAVITY-DEPENDENT PHENOMENA IN PHYSICAL SCIENCES, Berlin, Germany, 2-5 May 1992 [1] and extensions therof [2].

With the advent of space flight and the development of relevant fluid systems, the dynamics of rotating fluids experienced anew the interest of engineers and scientists. The behavior of spin-stabilized rockets and satellites is only one example. Under the conditions of weightlessness or microgravity (μ-g) further applications of such systems are conceivable, e.g. the management of multiphase systems in spherical or cylindrical configurations, or the containerless handling of rotating liquid columns in connection with the semiconductor-single crystal fabrication by the Czochralski method. By rotation of the appropriate configuration it is possible to apply acceleration fields systematically, which again is desirable with respect to some phases of a production process in orbit.

In the present study ground-based experiments were carried out to determine the natural modes and the natural frequencies of a two-phase immiscible fluid system (petrolether 40/60 and water), enclosed in a rapidly rotating cylindrical vessel subject to an axial excitation. The undisturbed cylindrical interface, being parallel to the direction of gravity g (9,81 m/s^2), is generated by a very high rotational frequency $f_d = \Omega/2\pi$, with Ω as the speed of rotation (angular velocity) about the spin axis (axis of rotation). Due to an axial excitation with a frequency $f_a = \omega/2\pi$, where ω denotes the natural frequency, a disturbance of the interface occurs. The resulting natural modes are documented photographically. The natural frequencies of the investigated liquid-liquid system depend on several parameters: the diameter D(=2R) of the vessel (27.7 mm and 32.7 mm), where R is the radius, the geometric ratio H/D (2.0; 1.5; 1.0; 0.5), where H is the height of the container, the filling ratio ε (0.5 and 0.75), the rotational frequency of the vessel (20; 40; 60 Hz), and the gradually adjustable axial excitation frequency (range 5 Hz to 300 Hz). The choice of the values of the filling ratio is based on the following consideration. In case of $\varepsilon = 0.5$, both liquids occupy the same volume. For $\varepsilon = 0.75$, the mean radius $a = R\sqrt{1-\varepsilon}$ of the interface from the tank axis is equal to half of the inner radius of the vessel.

2. Experimental

Figure 1 displays schematically the immiscible liquid-liquid system in a rotating cylindrical vessel, a photograph of the test stand is shown in Figure 2. For a description of the experimental setup and the performance of the experiments reference is made to [3]. If the container, rotating with a high angular velocity, is excited additionally in such a way that the rotational frequency $\Omega = 2\pi f_d$ corresponds to a natural frequency $\omega = 2\pi f_a$, the undisturbed cylindrical interface is disturbed. Due to the observation technique the interface appears as a standing wave.

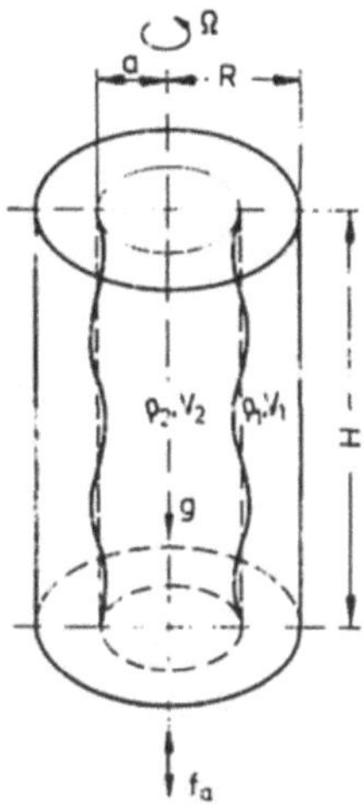

Figure 1: Arrangement of two immiscible liquids in a rapidly rotating cylindrical container, geometry and notation

Figure 2: Photograph of the test stand

For reasons of symmetry and continuity, only the half wave trains of the odd eigenvalues of the configuration are observable [4]. The order of the eigenmodes EM is determined by the number of the vibration nodes of the interface wave. The wavelength λ of the eigenmode is given by

$$\frac{\lambda}{2} = \frac{H}{EM} . \qquad (1)$$

For an intuitive derivation of this equation by means of an analogy model (transverse vibrations of a string fixed at both ends), see [5]. Figures 3 and 4 exhibit examples of observed eigenmodes. The pictures shown were used either as photographs or as negatives for the quantitative interpretation of the experiments by counting the number of the vibration nodes. Between the relevant parameters we assume a relation

$$\varphi\,(a, T_{12}, a\Omega^2, \rho_1, \rho_2, g, H; \omega) = 0\,, \qquad (2)$$

with T_{12} ($25.6 \cdot 10^{-3}$ N/m) as the interfacial tension between the two liquids, ρ_1,ρ_2 ($> \rho_1$) as the liquid densities for petrolether 40/60 (654 kg/m^3) and water (998 kg/m^3), and $a\Omega^2$ as the centrifugal acceleration. An UV-selective dye-stuff (Uranin) was added to the water phase. The interfacial tension was determined by a ring-tensiometer.

Figure 3: 7th Eigenmode

Figure 4: 15th Eigenmode

The dimensional matrix

	a	T_{12}	$a\Omega^2$	ρ_1	ρ_2	g	H	ω
length	1	0	1	-3	-3	1	1	0
mass	0	1	0	1	1	0	0	0
time	0	-2	-2	0	0	-2	0	-1

is of the third rank. According to the general Pi Theorem of Vaschy and Buckingham there follows

$$\psi\,(\Pi_1, \ldots, \Pi_5) = 0\,. \tag{3}$$

The density ratio $\Pi_4 = \rho_2/\rho_1$ remains constant, and thus yields no contribution. Likewise $\Pi_5 = g/\,a\Omega^2$ (Froude number) can be neglected also, since $(a\Omega^2/g) >> 1$. Since ρ_1, ρ_2 and T_{12} are fixed numbers, the following Π-terms are important with respect to the experiments

$$\Pi_1 = Bo = \frac{(\rho_2 - \rho_1)\,a^3\,\Omega^2}{T_{12}} = F_1\,(f_d, D, \varepsilon)\,, \tag{4}$$

$$\Pi_2 = \frac{H}{a} = F_2\,(H, D, \varepsilon)\,, \tag{5}$$

$$\Pi_3 = \frac{f_a}{2f_a} = F_3\,(f_a, f_d)\,, \tag{6}$$

where Bo denotes the Bond number. This non-dimensional parameter compares the capillary force to the volume force. Finally, an additional dimensionless number occurs in the course of the experiments, namely the filling ratio

$$\varepsilon = \frac{V_1}{V_1 + V_2}\,, \tag{7}$$

with V_1 as the Volume of the petrolether 40/60 and V_2 as the volume of the water. However, this number remains constant for the respective experiments.

3. Discussion of selected test results

Figures 5-7 show test results demonstrating the dependence of the eigenmode EM from the dimensionless frequency ratio $f_a/2f_d$. The filling ratio and the geometric ratio are kept constant. Inspection of the plots reveals that they can be divided into two sections:

1. This section holds for the range$(f_a/2f_d) < 1$.

   For the filling ratio $\varepsilon = 0.75$ and the geometric ratio $(H/D) = 2$ the section extends up to the 13th eigenmode. In this diagram the curves for different rotational frequencies exhibit the same behaviour.

2. This section holds for the range $(f_a/2f_d) > 1$.

   For the filling ratio $\varepsilon = 0.75$ and the geometric ratio $(H/D) = 2$, the section starts with the 13th eigenmode. One clearly recognizes different courses for different rotational frequencies. The higher the vibrations per minute (v.p.m.), $n = 30\,\pi/\Omega$, the less the dimensionless frequency to generate a constant eigenmode. The distance of the curves with different v.p.m. increases continually from the 13th eigenmnode. For a constant eigenmode, the difference between the two curves decreases with increasing excitation frequency. The two distinct ranges in the figures 5-7 can be attributed to different centrifugal forces. The latter increase with increasing rotational frequency.

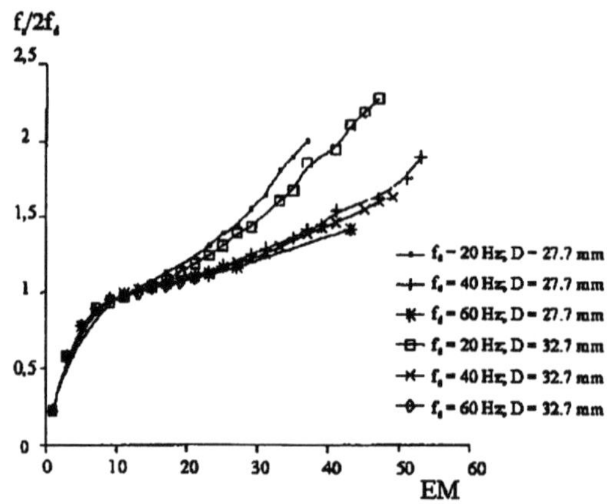

Figure 5: Plot of $f_a/2f_d$ vs. EM; ε = 0.75; H/D = 2.0

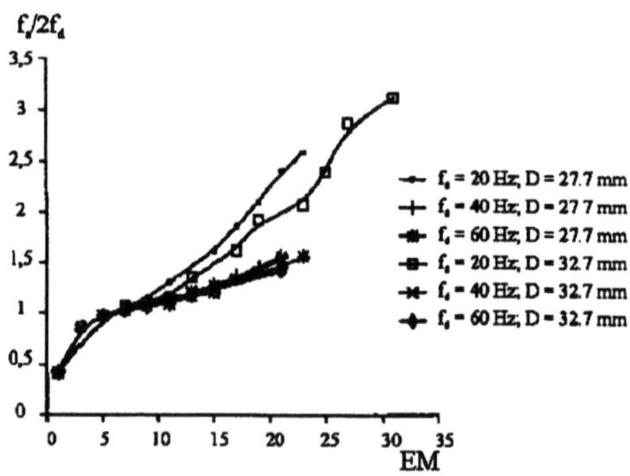

Figure 6: Plot of $f_a/2f_d$ vs. EM; ε = 0.75; H/D = 1.0

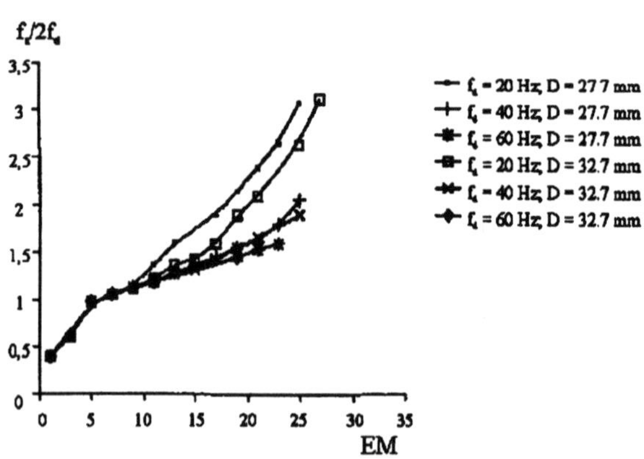

Figure 7: Plot of $f_a/2f_d$ vs. EM; ε = 0.5; H/D = 1.0

In order to generate a distinct eigenmode, the excitation frequency has to overcome the stability of the initial position (liquids without eigenvibrations), due to the centrifugal force. Thus to get such a distinct EM, higher excitation frequencies are required for higher rotational frequencies. Of interest is the superposition of the curves in section 1 of the dimensionless representation. This means that there exists a linear dependence between the excitation frequency and the rotational frequency. This does not hold, however, in section 2, since the curves do not exhibit the same course.

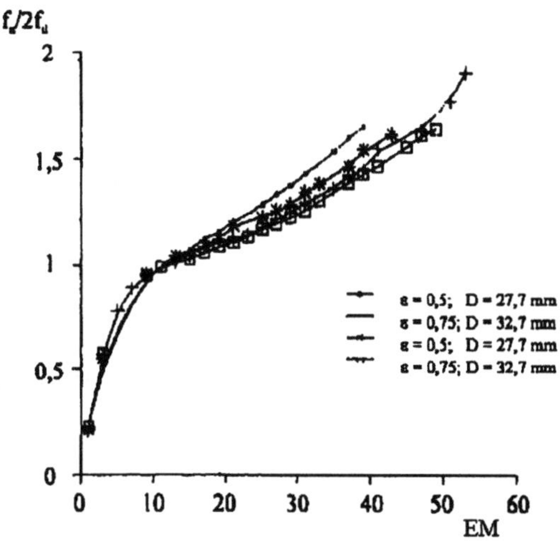

Figure 8: Plot of $f_a/2f_d$ vs. EM; H/D = 2.0; f_d = 40 Hz

Figure 8 displays the dependence of the eigenmode from the filling ratio. The dimensionless frequency ($f_a/2f_d$) is plotted vs. the eigenmode EM. The curves are almost congruent, only in the range ($f_a/2f_d$) > 1 a disturbance is noticeable. In spite of the small deviation one recognizes the dependence from the filling ratio. The larger the filling ratio, the smaller the excitation frequency to reach a distinct eigenmode.

Figure 9 shows a plot of the dimensionless frequency ($f_a/2f_d$) vs. the Bond number for different filling ratios. One distinguishes three types of curves:

1. Eigenmodes for which ($f_a/2f_d$) < 1 holds.

   The curves for $\varepsilon = 0.5$ and $\varepsilon = 0.75$ are different for the same eigenmodes. In case of a constant eigenmode the required dimensionless frequency is smaller for $\varepsilon = 0.5$ than $\varepsilon = 0.75$.

2. The transition range ($f_a/2f_d$) ≈ 1.

   The curve of the different filling ratios are superimposed.

3. Eigenmodes for which $(f_a/2f_d) > 0$ holds.

   In order to get a definite eigenform one observes that the dimensionless frequencies are lower for curves with $\varepsilon = 0.75$.

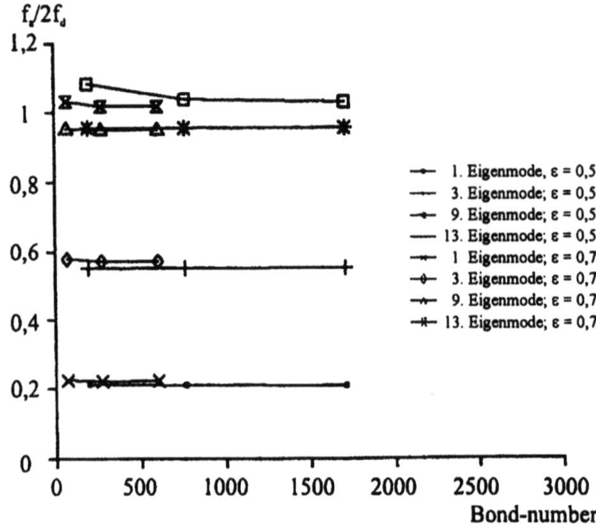

Figure 9: Plot of $f_a/2f_d$ vs. Bond-number; H/D = 2.0; D = 27.7 mm

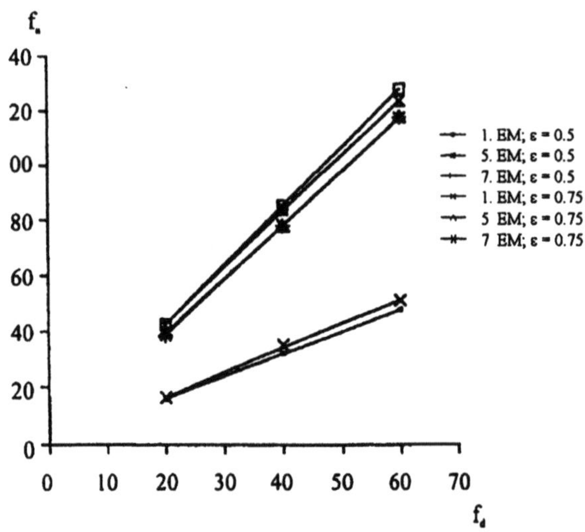

Figure 10: Plot of f_a vs. f_d; H/D = 1.0; D = 27.7 mm

For $\varepsilon = 0.75$ the fraction of the heavy liquid in the tank is about 25 % larger than for $\varepsilon = 0.5$. This means that in the rotating steady state a higher centrifugal force is acting perpendicular to the tank axis. With that there exists at the start (for small eigenmodes) a larger resistance force to generate an eigenmode for the liquid with $\varepsilon = 0.75$ than for the case $\varepsilon = 0.5$. For $\varepsilon = 0.75$ a higher dimensionless frequency occurs for a constant low eigenmode.

It turns out that the low eigenmodes are stable only for a very narrow range of the excitation frequencies. The higher the eigenmodes, the larger the range of the excitation frequency. Along with the transition to the next eigenmode the liquid is always in a state of weak vibrations.

Figure 10 depicts a plot of the excitation frequency vs. the rotational frequency for different filling ratios, if the geometric ratio and the diameter are kept constant. Type 1 holds for very low eigenmodes (in the figure the first EM). For the same rotational frequency the excitation frequency is higher for $\varepsilon = 0.75$ to reach the same eigenmode as in case of $\varepsilon = 0.5$. Type 2 characterizes the domain of superposition of the curves for different filling ratios. The figure shows this for the fifth eigenmode. Finally, Type 3 is remarkable by the fact that for a constant eigenmode the filling ratio $\varepsilon = 0.5$ leads to a higher excitation frequency as $\varepsilon = 0.75$.

Acknowledgement

The authors would like to take this opportunity to thank Ms. Petra Zapp and Mr. Thomas Eich for their assistance by the performance and evaluation of the experiments

References

[1] Siekmann, J. and Raake, D.: Experimental investigation of the natural modes of immiscible liquids enclosed in an axially excited rapidly rotating vessel. *Abstracts Ninth European Symposium* GRAVITY-DEPENDENT PHENOMENA IN PHYSICAL SCIENCES, Berlin, Germany, 2-5 May 1995, pp. 315-316, DARA, TU Berlin, ESA, Symposium Organizer Professor M.J. Fecht, Technical University Berlin (1995).

[2] Siekmann, J. and Raake, D.: Experimentelle Untersuchung von Eigenschwingungsformen des Stoffsystems Petrolether 40/60 und Wasser in einer axial angeregten, schnell rotierenden Küvette. Festschrift zum 60. Geburtstag von Prof. Dr.-Ing. Reint de Boer, *Beiträge zur Mechanik, Forschungsbericht aus dem Fachbereich Bauwesen 66*, Universität-GH Essen, Dezember 1995, S. 367-377, 1995.

[3] Raake, D.: Untersuchung axial erregter Eigenschwingungen nichtmischbarer Fluide in einer schnell rotierenden zylindrischen Küvette. *Dissertation* Universität-GH Essen, 1991

[4] Raake, D. and Siekmann, J.: Axially excited natural frequencies in a rapidly rotating cylindrical container. *Z. Flugwiss. Weltraumforsch.* 15 (1991), 289-296.

[5] Himmel, A., Raake, D. und Siekmann, J.: Experimentelle Untersuchung von Eigenschwingungsformen nichtmischbarer Fluide in einer zylindrischen, schnell rotierenden, axial angeregten Küvette. *Forschung im Ingenieurwesen-Engineering Research* Bd. 61 (1995) Nr. 1/2, S. 6-11.

Routes into Chaos in Rotating Fluid Flows

Christoph Egbers and Hans J. Rath

Center of Applied Space Technology and Microgravity (ZARM),
University of Bremen, 28359 Bremen, F.R.G.

Abstract. We report on a concurrent study on stability, bifurcation scenarios and routes into chaos in rotating fluids, i.e. Taylor Couette flow, spherical Couette flow and rotating annulus heated from within (Rossby tank). In contrast to the pitchfork bifurcation as the first instability in the cylindrical system the flow in the spherical gap flow bifurcates via different Hopf bifurcations into chaos. By increasing the Reynolds number with the angular velocity of the driving inner sphere the flow bifurcates from laminar axisymmetric basic flow to the periodic motion of non-axisymmetric spiral waves for relative large aspect ratios. The spiral waves exist over a wide range of the Reynolds number. In this range a change in shape and periodicity can be detected by visualization with small aluminum flakes and also measured by Laser Doppler velocimetry (LDV-technique). At high Reynolds numbers, the flow undergoes a bifurcation to low-dimensional chaotic motion before it becomes turbulent. The dynamic behaviour of the rotating flows mentioned above is compared and discussed by time series analysis methods and by velocity bifurcation diagrams.

1 Introduction

The subject of hydrodynamic instabilities and the transition to turbulence has an importance for the understanding of non-linear dynamic systems. Progress in understanding instabilities, bifurcations and routes into chaos has been made primarily by focusing attention on a small number of relatively simple hydrodynamic systems like Rayleigh-Be'nard convection and the flow between two concentric rotating cylinders (Taylor-Couette flow). Furthermore, a considerable progress in understanding the first instability in the form of Taylor vortices of a viscous incompressible fluid flow between two concentric rotating spheres for small and medium gap widths has been achieved over the last two decades. The three examples just reviewed are examples of transition to turbulence through a repeated finite number of symmetry-breaking bifurcations. Especially the study of instabilities and turbulence in spherical Couette flow and in the rotating annulus is of basic importance for the understanding of global astrophysical and geophysical motions. Much of the universe is filled with fluids in turbulent motion, and instabilities are quite common in planetary atmospheres. But the study of spherical Couette flow is also important for general theory of hydrodynamic stability since this flow is a natural combination of circular Couette flow at the equator and the flow between rotating disks at the poles. Another important feature of the spherical geometry is that the basic flow involves two types of symmetry, the reflection symmetry with respect to the equator and the translational symmetry with respect to the axis of rotation. Depending on the aspect ratio, both types of symmetry-breaking bifurcations can exist in the spherical Couette flow.

In this work, we consider three different types of rotating fluid flows: The flow between two concentric spheres, the flow between two concentric cylinders (Taylor Couette) and the flow in a cylindrical annulus heated from within (Rossby tank) as illustrated in figure 1. As an example some typical instabilities occuring in these systems are depicted in figure 2.

Figure 1: Photographs of the rotating systems

Figure 2: Pictures of flow visualization studies in rotating systems

Most of previous experimental investigations on the spherical Couette flow were restricted to small and medium gap widths, where Taylor vortices exist as reported by Sawatzki & Zierep (1970), Munson & Menguturk (1975), Yavorskaya et al. (1980), Wimmer (1981), Bühler (1985) and Bühler & Zierep (1984, 1986). Although spherical Couette flow is more relevant to astrophysical and geophysical applications, in comparison with the Taylor-Couette flow system (e.g. Fenstermacher et al., 1979, Andereck et al., 1986, Buzug et al., 1992, 1993, v. Stamm et al., 1993, Mullin, 1993) the dynamic behaviour during the laminar-turbulent transition of the flow between two concentric spheres has been studied less (e.g. Belyaev et al., 1984, Nakabayashi & Tsuchida, 1988, Egbers & Rath, 1995).

Some new aspects of the dynamic behaviour of the spherical Couette flow during the transition to turbulence for a wide range of Reynolds numbers and for the case of the wide gap width ($\beta = 0.5$) are presented in this study. The bifurcation scenario arising in the spherical Couette system is in contrast to the bifurcation scenario in the Taylor-Couette system as can be seen from our results described in chapter 6. The flow in the wide spherical Couette flow for example bifurcates via different Hopf-bifurcations. The instabilities occuring are in the form of non-axisymmetric spiral waves, which break the spatial symmetry-behaviour of the basic flow. With increasing the Reynolds number, the number of spiral waves decreases, before the flow loses its stability and it seems to become chaotic with increasing the Reynolds number (Egbers & Rath, 1995, Wulf et al, 1997). This transition was earlier thought to be a direct transition into turbulence without the existence of instabilities (e.g. Munson & Menguturk, 1975).

With the means of time series analysis of the LDV-measurements, we want to give some additional quantitative estimates for a comparison of the different routes into chaos in rotating systems. Experimental results connected with the problems of stability, bifurcation, non-axisymmetry, periodicity, quasi-periodicity, chaotic and turbulent motions in rotating fluid flow systems will be presented.

Two theoretical models of the transition from a laminar basic state to the turbulent one are summarized in chapter 2. The experimental setups used for investigation are described in chapter 3. The LDV-measuring system is described in chapter 4, while the time series analysis methods are explained in chapter 5. Experimental results on simultaneous flow visualization studies and the LDV-measurements on the dynamic behaviour during the laminar turbulent transition in rotating systems are presented in chapter 6. In chapter 7 we give some conclusions.

2 Theoretical Models of the Transition to Turbulence

In this chapter, we summarize previous theoretical models of the transition from laminar basic to turbulent flows: The first model was postulated by Landau (1944) presenting the following idea. If the fluid flow is in some laminar state and the stress parameter - the Reynolds number - is increased then a simple periodic oscillation will be excited in the flow above a critical value of the Reynolds number. Further increase in Reynolds excites another oscillation whose frequency is incommensurate with the first one and the process continues so that more and more modes are excited sequentially as the Reynolds number is increased. Each of the new modes is introduced at smaller and smaller increments of the Reynolds number so that an accumulation point is rapidly reached where a continuum of discrete modes is excited. In Landau's model, turbulence may be thought of in these terms as an infinite sum of discrete incommensurate frequencies with random phase. These ideas are consistent with a statistical description of turbulence, which has been successfully used in the development of turbulence models.

More recently, Ruelle & Takens (1971) suggested that the laminar basic state can be thought of as an attracting fixed point in the phase space. Thus, if the system is disturbed from this state, then any resulting trajectory will head back towards the fixed point. If the Reynolds number is increased in this case, a time-periodic mode will be excited in the flow which can be represented by a limit cycle in phase space. A further increase in the Reynolds number will lead to the appearance of a second frequency which is incommensurate with the first one. This will give a torus in phase space. As the Reynolds number is increased furthermore, a third frequency begins to appear which would form a three frequency torus. Ruelle & Takens (1971) suggested, however, that this is a non-robust situation and it is more typical for a so-called strange attractor to be in the solution structure at this stage. This model is said to be a typical description for classes of equations for which the Navier-Stokes equations are believed to belong. This has not yet been proved for all fluid-dynamic problems but evidence is mounting that, at least in certain confined flows - as i.e. Taylor Couette flow is - it might be a better descriptor than the Landau picture. But the question is, which bifurcation scenario can describe the route into chaos in the spherical Couette flow in contrast to Taylor Couette flow or rotating annulus heated from within and which model is suited to describe this transition?

3 Experimental Apparatus

A universal rotating table was constructed to adapt the different rotating systems, i.e. the cylindrical, the spherical and the rotating annulus. Generally, both cylinders (spheres) can be rotated independently by means of two belt-drives. The eccentricity between outer and inner cylinder (sphere) could be minimized to ±0.015 mm using high precision bearings and shafts. Two different synchronous motor drives were adapted, which provide a uniform and stable rate of rotation up to n = 850 rev/min with fluctuations of less than 1.5%. They variate the Reynolds numbers of the inner and outer cylinder (sphere) respectively in a range from 0 to Re = 10^6 (10^5). The revolutions were calibrated using an optical tracking system (optical coupling). Most of our experiments were carried out by increasing the Reynolds number quasistationary from zero. However, because the occuring flow structures depend also on initial conditions (Egbers 1994), the acceleration rate for both cylinders (spheres) could be variated. Experiments with the rotating flow systems were carried out in a laboratory condition, where the temperature could be kept uniformly up to ± 0.3°C. Since the viscosity of the silicone oils, which were used as working fluids, vary by approximately 2% / ^{0}C, the temperature must be precisely controlled and measured in order to have a well defined Reynolds number. A temperature accuracy of ± 0.15^{0}C was achieved for the temperature sensors (PT 1000) used in the experiments.

3.1 Spherical Couette Flow Experiment

The spherical Couette flow can be characterized by the following control parameters: The aspect-ratio $ß = (R_2 - R_1)/R_1$ and the Reynolds number $Re = (R_1^2\ \Omega)/\nu$, where R_1 and R_2 are the inner and outer radii, Ω is the angular velocity of the inner sphere and ν is the kinematic viscosity. Another control-parameter coming into account is the acceleration rate $d\Omega/dt$, where t is time, because the occuring flow pattern during the transition to turbulence are also determined by the history of the flow, i.e. it depends on whether the Reynolds number is increased or decreased, quasistationary or fast.

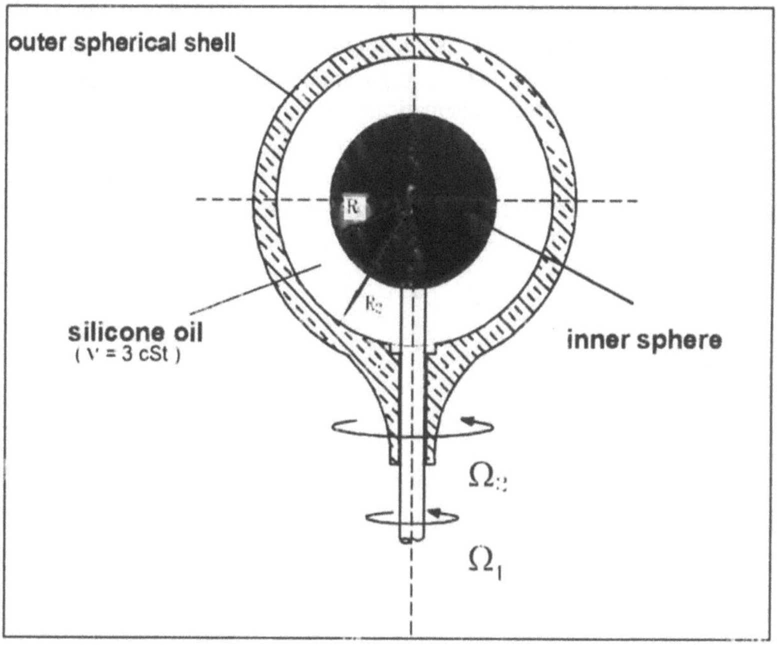

Figure 3: Principle sketch of the spherical Couette flow model

The experimental setup for the spherical Couette flow consists of an inner sphere rotating concentrically inside another rotating outer spherical shell. A principle sketch of the experimental model is illustrated in figure 3: The outer sphere (R_2 =40.00±0.02 mm) is composed of two transparent acrylic plastic hemispheres. The inner spheres are made out of aluminium having the five various radii R_1, to variate the aspect ratio ß from 0.08, 0.18, 0.25, 0.33 to 0.5.

For measuring the fluid temperature, three temperature sensors are installed just below the outer surface of the inner sphere at latitudes $\Theta = 10^0$, $\Theta = 80^0$ and $\Theta = 160^0$. These temperature data were transferred from the rotating sphere to the stationary part by a slip-ring system. For investigations with only the inner sphere rotating three other temperature sensors are installed at the inner surface of the outer sphere at latitudes $\Theta = 0^0$, $\Theta = 45^0$ and $\Theta = 90^0$. Our measurements were carried out with the silicone oil M3 at a constant temperature of 25^0 C. As tracer particles for flow visualization studies small aluminium flakes were used. The concentration by weight was 0.05%. The viscosities of the working fluids with tracer particles were measured with a VOR-rheometer (BOHLIN-Reologi AB, Sweden). An effect on the viscosity was not detected.

3.2 Taylor Couette Flow Experiment

The Taylor Couette flow (Taylor, 1923) is characterized by the following three control parameters: The aspect-ratio $\Gamma = H/d$, the radius ratio $\eta = R_1 / R_2$ and the Reynolds number $Re = (R_1\, d\, \Omega)/\nu$, where R_1 and R_2 are the inner and outer radii, d is the gap width, H is the height, Ω is the angular velocity of the inner cylinder and ν is the kinematic viscosity. The experimental setup with the applied LDV-measuring system is illustrated schematically in figure 4. The inner cylinders are made out of aluminium having six different radii R_1, to variate the radius ratio η from 0.5, 0.67, 0.75, 0.8, 0.85 up to 0.925. Both cylinders can be rotated independently. The end-plates are fixed. In this work we used a small ($\eta = 0.85$) and a wide cylindrical gap ($\eta = 0.5$) for the investigations on bifurcation scenarios.

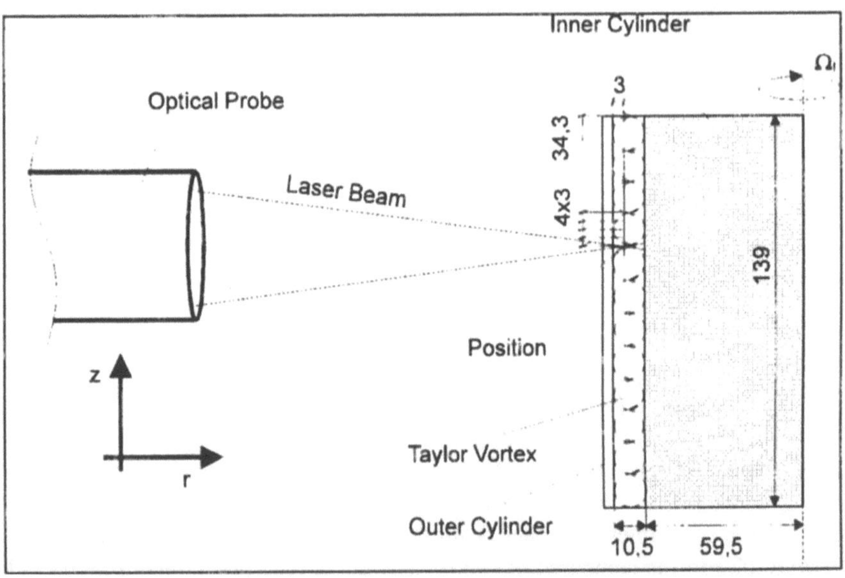

Figure 4: Principle sketch of the Taylor Couette system with the applied LDV technique

3.3 Rotating Annulus Heated from within

Rotating annulus experiments (Hide & Mason, 1975) have been carried out primarily to obtain laboratory simulations of the large-scale circulation of the atmosphere and recently for vortices (Marcus, 1988), such as Jupiter's Red Spot.

The flow in the rotating annulus heated from within can be characterized by the following control parameters: The geometry can be described by the radius-ratio $\eta = R_1 / R_2$ and the aspect ratio $\Gamma = H/d$, where R_1 and R_2 are the inner and outer radii, H is the height and d is the gap width of the annulus. Furthermore, the Taylor number $Ta = \frac{4 \cdot \Omega^2 \cdot d^5}{\nu^2 \cdot H}$ and the Rossby number $Ro = \frac{g \cdot H \cdot \Delta\rho}{\bar{\rho} \cdot \Omega^2 \cdot d^2}$ are coming into account, where Ω is the angular velocity, ν is the kinematic viscosity and ρ is the density. The Taylor number describes the rotational influence while the Rossby number comprises the influence of density gradient due to temperature differences and the influence of rotation. In Figure 5 the experimental setup of the rotating annulus is illustrated. The inner cylinder can be heated and the outer one can be cooled by external temperature baths. The temperature difference between inner and outer cylinders is controlled by temperature sensors installed just below the surfaces of inner and outer cylinders. The fluid motion on the free surface of the cylinders is visualized by aluminium flakes suspended in the working fluid, which is in this case water.

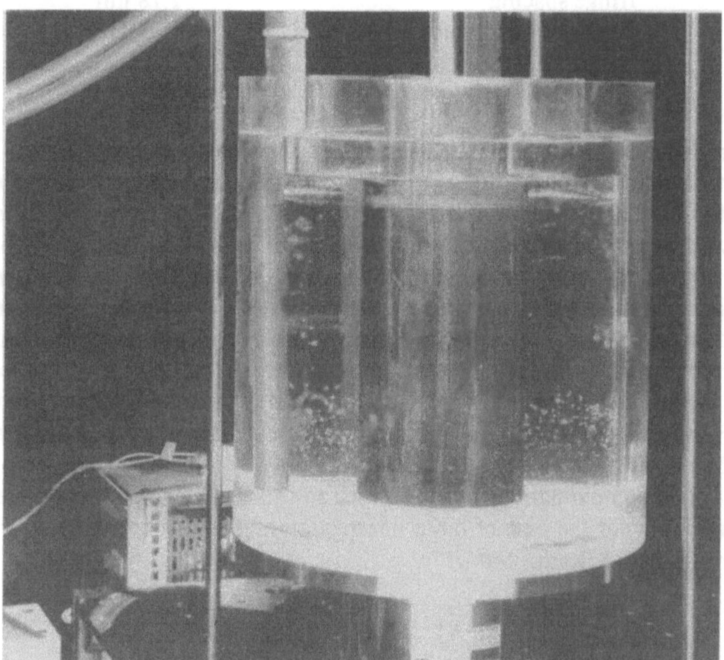

Figure 5: Photograph of the rotating annulus experiment

4 Measuring Techniques

For the application of the laser-Doppler-velocimetry (LDV) on the rotating systems, special traversing systems have been constructed. As an example the applied LDV technique with the mounted LDV optic probe on the spherical Couette flow experiment is depicted in figure 6. The laser-Doppler-velocimeter system used in our experiments, consists of a 10 mW He-Ne-laser and a 1-D fiber flow optic probe (DANTEC-Electronics, Denmark). A frequency shift is added by the Bragg-cell to one of the beam pair to allow for measurements of reversing flows. The optic probe with a fibre optic cable is connected to the laser via manipulators. The backscattered light is focused on a photomultiplier tube. Data records of each velocity measurement consisted of 32000 validated Doppler bursts. Parameters characterizing the LDV-system are listed in table 1.

He-Ne laser	10 mW
focal length	160 mm
wavelength	632.8 nm
beam spacing	38 mm
beam diameter	1 mm
number of fringes	22
probe volume:	
length	0.6 mm
diameter	0.07mm
fringe spacing	2.18 μm
Bragg-cell frequency	40 MHz

Table 1: Parameters characterizing the LDV-measuring system

As tracer particles for the LDV-measurements polysterene spheres (Bangs-Lab., USA) with a diameter of 1.6 μm were used. The concentration of tracers in the working fluid was 0.01 Vol.%. The application of the LDV-technique on the spherical as well as on the cylindrical annulus requires an optical correction for the accurate determinations of the probe volume locations and for the interference fringe spacing due to refraction effects of the curved outer surface. Because the probe is adjusted in radial direction, the optical axis of the front lense of the probe passes perpendicular through the curved outer surface. Thus, the correction for the two laser beams, which are in the same plane, could be calculated for a cylindrical surface. However, the fact, that a small probing volume is needed to produce sufficient spatial resolution, which could be obtained only by a large intersection angle, the small-angle approximation cannot be used in this case. The correction method used in this work for the case of large intersection angles was derived in our previous work (Egbers & Rath, 1996).

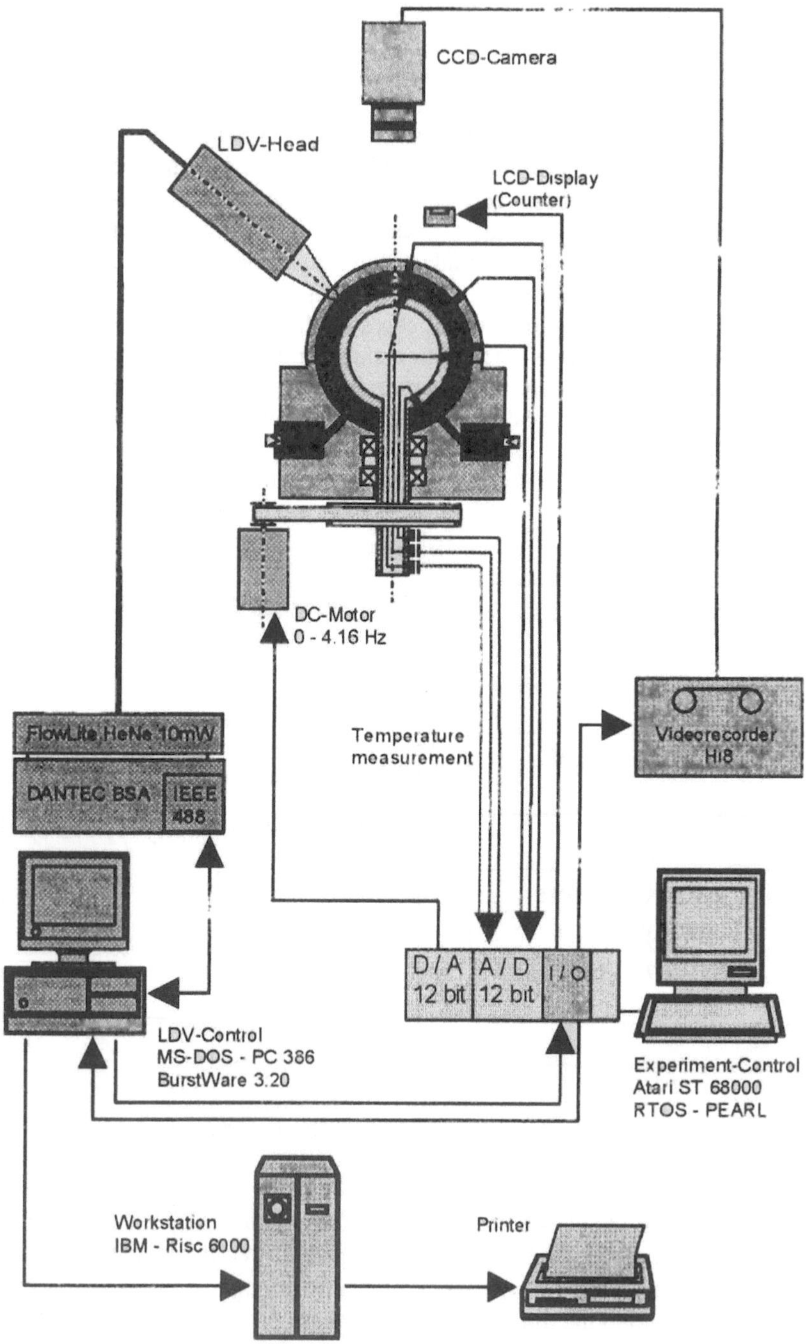

Figure 6: Principle sketch of the applied LDV measuring system on the spherical Couette flow experiment

5 Methods of Time Series Analysis

Time series analysis methods of non-linear systems have been improved over the recent decade leading to a much better understanding of the dynamics occuring on the route to chaos in different fluid systems. First investigations by Gollub and Swinney (1975) using Fourier analysis of LDV time series confirmed sucessfully the Rouelle-Takens-Newhouse scenario (1978) for the onset of turbulence in the Taylor Couette flow. However, linear methods are limited for the study of chaotic states. Especially for the case that fundamental frequencies are still remaining in a week chaotic flow a classification of the motion is hardly possible. This problem has been overcome for low dimensional chaos by the reconstruction method of the underlying attractor in phase space from a single time series containing characteristic information of the appearing dynamics. In this chapter, we introduce the pre-processing of data, the calculation of bifurcation diagrams using linear methods and the analysing techniques for reconstructed space. A principle sketch of the time series analysis using the LDV-technique is illustrated in fig. 7.

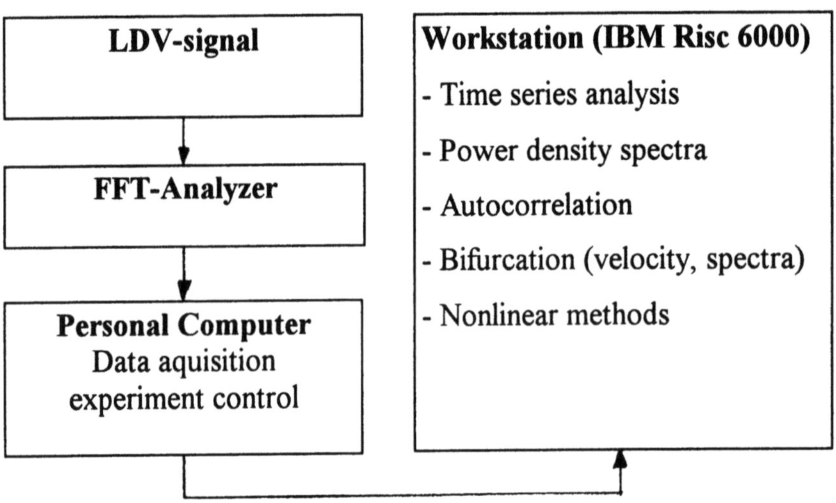

Figure 7: Principle sketch of the LDV data analysis

The pre-processing includes a resampling of the obtained LDV time series by a linear approximation between two collected signals to recieve a constant time increment Δt since the particles are passing randomly through the measuring volume. After the discretisation the time series are normalised on the time period of rotation and pre-filtered by a low-pass filter in the frequency domain with a cut-off frequency $f_{cut} = 6\ f_{rot}$ and a polynomial Savitzky-Golay filter (Press et al., 1994) with an order of four in the time domain. The chosen value of the cut-off frequency is high enough to preserve the significant frequencies of the system.

From the filtered time series the extrema are calculated by taking the first derivation from a polynomial approximation. The velocity values of the extrema are collected in histograms that are subsequently fitted with Gauss curves by a non-linear Levenberg Marquardt algorithm (Fechtmann et al., 1996) as illustrated in figure 8. The means of the Gauss curves are plotted against the Reynolds number yielding a representation of the dynamical changes of the flow. This type of representation is denoted as an extremal bifurcation diagram.

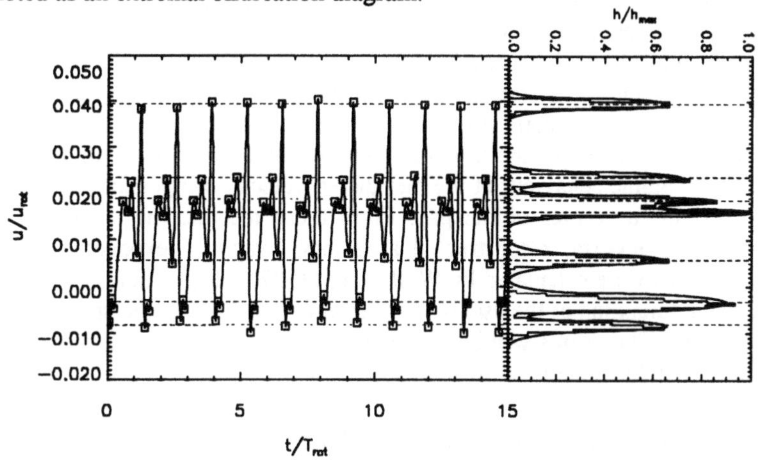

Figure 8: Determination of the significant extrema of a filtered time series for an extremal bifurcation diagram (Fechtmann et al., 1996)

As these methods are linear, they do not exhibit information on the distinct types of chaotic motion at higher Reynolds numbers. A technique going more into details of the chaotic motion is the reconstruction of the underlying attractor representing the dynamic behaviour. This is done by the method of time delayed coordinates introduced by Takens (1980) yielding state vectors representing the trajectories.

The Lyapunov spectrum of an attractor is calculated by the properties of the tangential flow mapping in phase space following the idea of Eckmann and Ruelle (1992). More details on the calculation of the time series is contained in the work of Fechtmann et al. (1996) and Wulf et al.(1997).

6 Results

6.1 Route into Chaos in the Spherical Couette Flow

In our previous investigation on the stability of spherical Couette flow (Egbers & Rath 1995), it was found, that the laminar basic flow between two concentric spheres, where only the inner sphere rotates and the outer one is at rest, loses its stability not only in the form of Taylor-instability. The instabilities occuring depend strongly on the aspect ratio. For wide gap widths ($\beta > 0.33$) considered here, the first instability manifests itself as a break of the spatial symmetry and a new non-axisymmetric spiral wave mode was observed, which spreads from the pole to the equator. With increasing the Reynolds number, the number of secondary waves with n spiral vortices decreases. For ß = 0.5 spiral waves with n = five, four or three spiral arms exist. Flow visualization studies on the development of the spiral wave flow up to chaotic motion are depicted in figure 10. The flow regimes, their spatial states with non-axisymmetric and non-equatorial symmetric secondary waves were characterized by measuring the phase velocities and the displacement between the northern and the southern hemisphere of the arising spiral waves (Egbers 1994). Furthermore, the spherical Couette flow in wide gap widths shows the well-known transition phenomena of hysteresis and non-uniqueness, depending on the acceleration rate of the inner sphere (Liu et al., 1996). In order to give some additional quantitative estimates on the dynamic behaviour of these spiral waves up to chaotic motions, the dynamic behaviour of the meridional velocity component (time series) was recorded in the center of the gap. From these records, the autocorrelation functions, power spectra, reconstructed attractors, Lyapunov exponents and the fratal dimensions were calculated by Wulf et al. (1997). Furthermore, by counting the minima and maxima of the oscillating velocities we obtained the spectral bifurcation diagrams. The results will be presented in the following chapters in comparison with the Taylor Couette bifurcation scenarios.

Figure 9 shows the evolution of the isothermal spherical Couette flow with an increasing Reynolds number for $\beta = 0.5$. The pictures were taken with a CCD-camera and digitised by a frame grabber facility. The difference of the gray values between a picture of an instability and a reference picture of the basic flow is calculated, amplified and pseudo coloured for each pixel to maintain only the structures of the instability. The first instability occurs in the form of five rotating axissymmetric spiral waves, then four spiral waves occur and finally three spiral waves which start to pulsate. This means that the waves begin to travel from their origin at the equator to the pole where they coincide and vanish. This state of three travelling spiral waves bifurcates again to five travelling waves with nearly the same dynamics appearing to the eye. The five waves change their shape into four travelling waves, which become asymmetric with respect to the pole and finally chaotic motion with large coherent structures is apparent.

The flow dynamics are obtained by LDV-technique with a sampling rate more then ten times higher then the highest significant frequency of the dynamic system. The obtained times series are filtered by an Optimal filter in the frequency domain to preserve the dynamic information from being eliminated by filtering (Fechtmann et al., 1996). An additional filtering is done by a Savitzky-Golay filter with an order of four. From the time series the power spectrum is calculated. The peaks in the power spectrum quantify the periodic behaviour of the flow. The frequencies of the peaks with a high significance are collected and presented in spectral bifurcation diagram as a function of the Reynolds number. The "traces" in the diagram give a good representation on how the motion changes from basic flow over different forms of periodic motion to chaos. The bifurcation scenarios are illustrated in the form of a velocity bifurcation diagram (fig. 10).

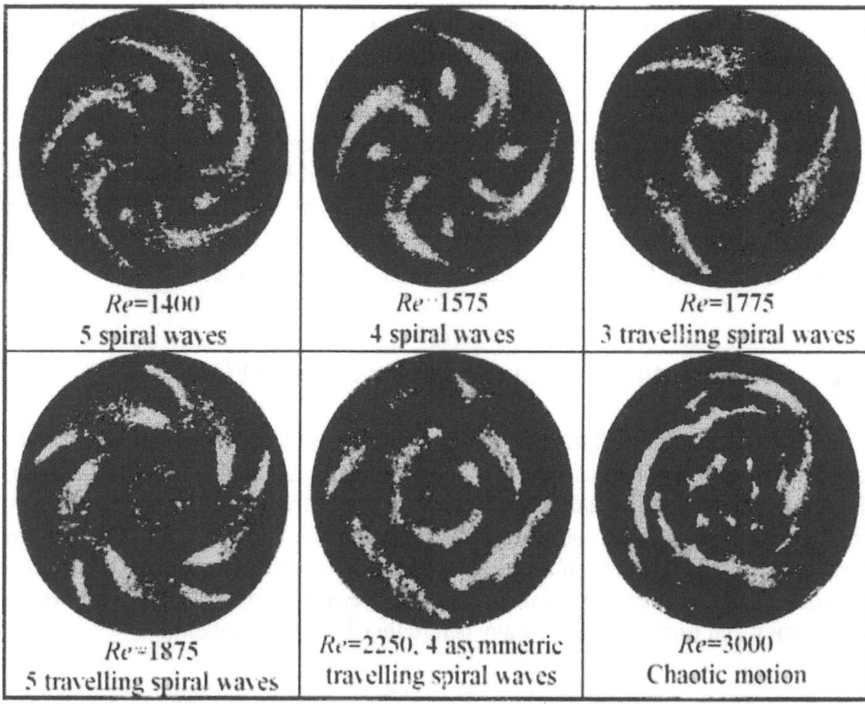

Figure 9: Flow structures of the wide gap instabilities for the gap width β=0.5

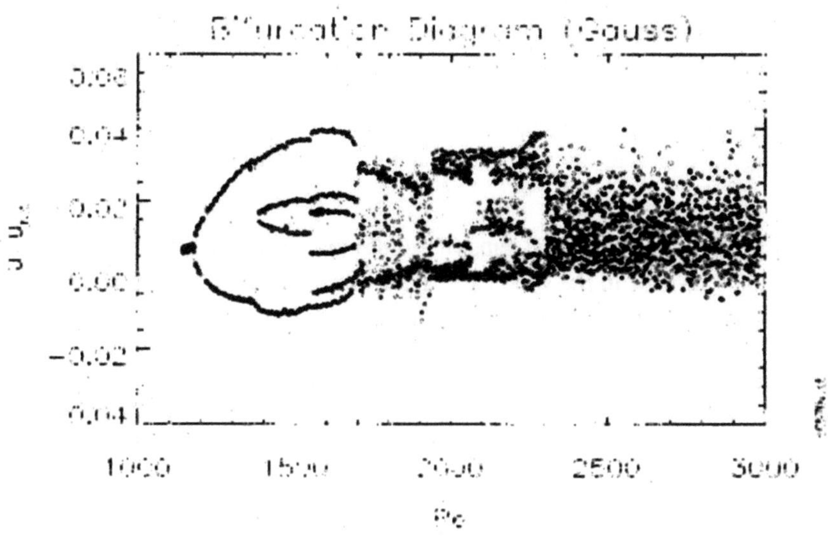

Figure 10: Bifurcation diagram for the spherical Couette flow at the position of $\Theta = 70°$ as a function of the Reynolds number. The meridional velocities are normalized by the velocity of rotation (ß = 0.5).

As can be seen from the diagram, the flow undergoes a bifurcation from laminar basic state to non-axisymmetric periodic motion in the form of five spiral waves at Re = 1170 (Wulf et al, 1997). The normalized frequency is $f_1 = 1.28$. At Re=1360 the shape of the periodic flow pattern changes to five rosettes with a normalized frequency of $f_1 = 1.02$. At Re=1550 four spiral waves exist. Finally the rosette modes decay to a spiral wave flow with three arms at Re=1700. This structure is also periodic with a normalized frequency of $f_1 = 1.28$ and begin pulsation at Re=1850. For Re ≥ 2080 only a week periodic motion (four travelling waves) can be observed by visualization. However, the bifurcation diagram shows that there is still periodic motion in the flow. The motion becomes clearly chaotic after Re ≥ 2300.

6.2 Route into Chaos in the Taylor Couette Flow

The experiments on bifurcation scenarios in the Taylor Couette system were carried out for two different radius ratios, $\eta = 0.85$ (small gap) and $\eta = 0.5$ (wide gap). Figure 11 shows flow visualization studies on the evolution of the isothermal Taylor Couette flow with an increasing Reynolds number for $\eta = 0.85$. The pictures were taken with a photographic camera. The first instability occurs in the form of parallel Taylor vortices. With increasing the Reynolds number these vortices become wavy and then modulated wavy before they become turbulent.

The flow dynamics are obtained by the LDV-technique in the same way as for the spherical Couette flow experiment. The velocity bifurcation diagram as illustrated in figure 12 was obtained as described in chapter 5 by collecting the extrema of the velocity time series with increasing quasi-stationary the Reynolds number. As can be seen from the bifurcation diagram, the flow undergoes a bifurcation from laminar basic state to a steady Taylor vortex flow via a pitchfork bifurcation, before the flow bifurcates via a Hopf bifurcation into the wavy mode. With a further increase of the Reynolds number the flow bifurcates via a second Hopf bifurcation into the modulated wavy mode, before chaotic motion occurs. The second critical Reynolds number for the onset of the wavy mode is of about 2.5 times of the first critical one.

In contrast to small cylindrical gaps ($\eta = 0.85$) just mentioned, the experiments on bifurcation scenarios in the wide gap Taylor Couette system $\eta = 0.5$ show another bifurcation scenario and another route into chaos. Flow visualization studies are depicted in Figure 13 and the bifurcation scenario is illustrated in figure 14. The basic flow seems to bifurcate via a Pitchforc bifurcation directly into chaotic motion. Periodicity or wavy modes were not found. The second critical Reynolds number for the onset of chaotic motion is of about 10 times of the first critical one.

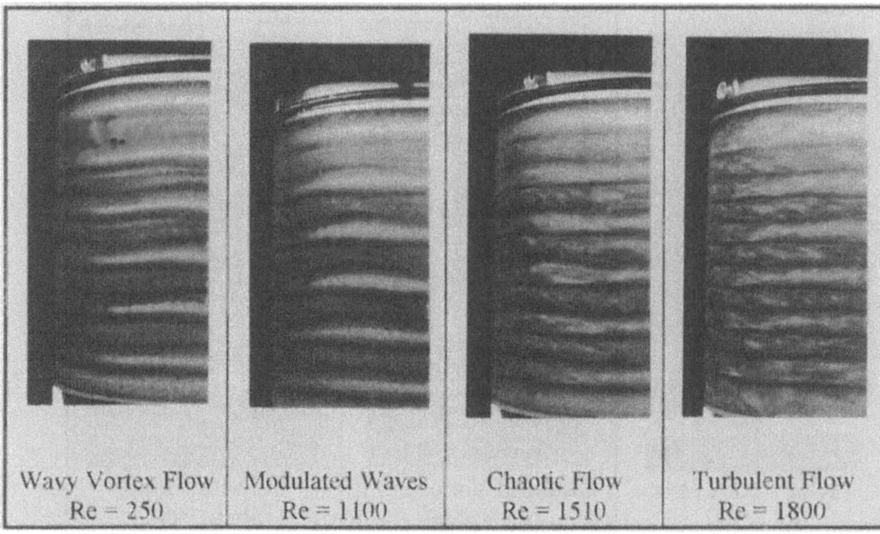

Figure 11: Flow visualization studies of the laminar turbulent transition in the Taylor Couette system (η=0.85, Γ=13.2)

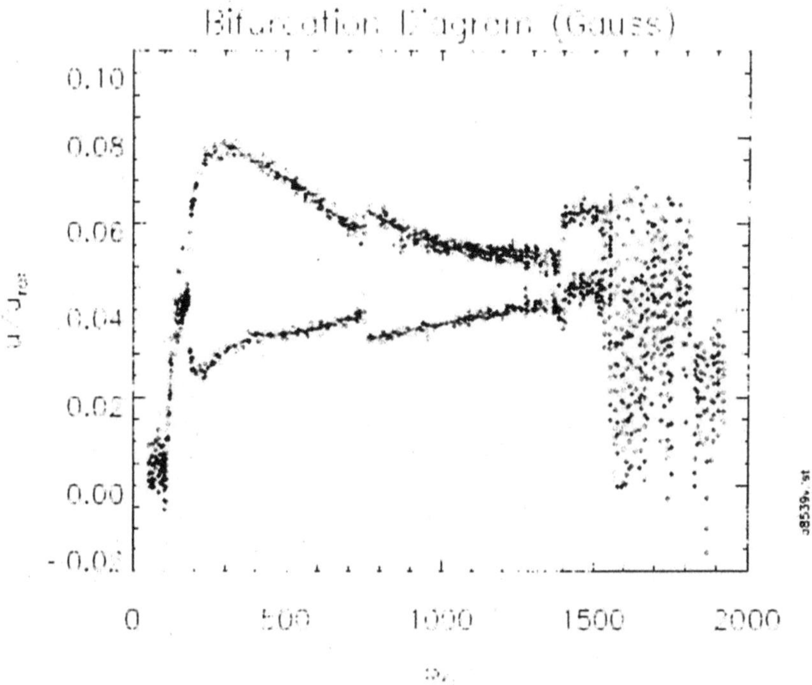

Figure 12: Bifurcation diagram in the Taylor Couette system as a function of the Reynolds number. The meridional velocities are normalized by the velocity of rotation (η=0.85, Γ=13.2, z=0.40H, $(r-R_1)/(R_2-R_1)$= 0.6)

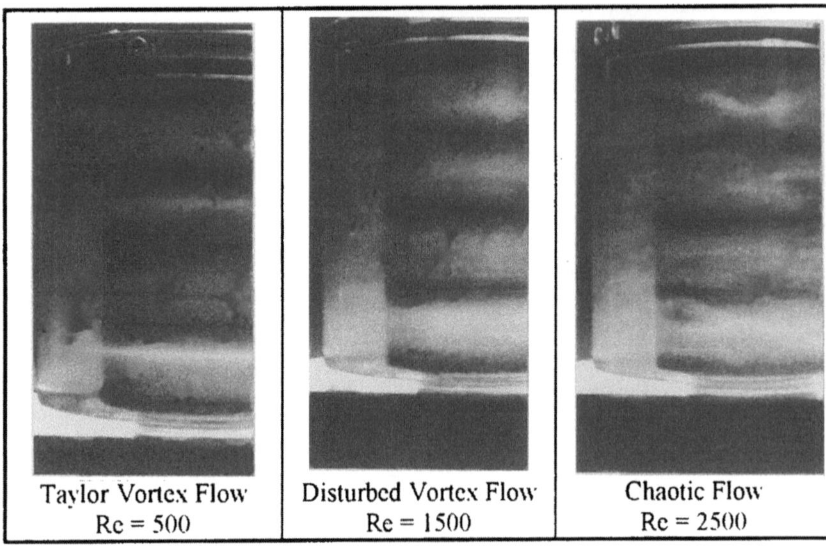

Figure 13: Flow visualization studies of the laminar turbulent transition in the Taylor Couette system (η=0.5, Γ=3.97)

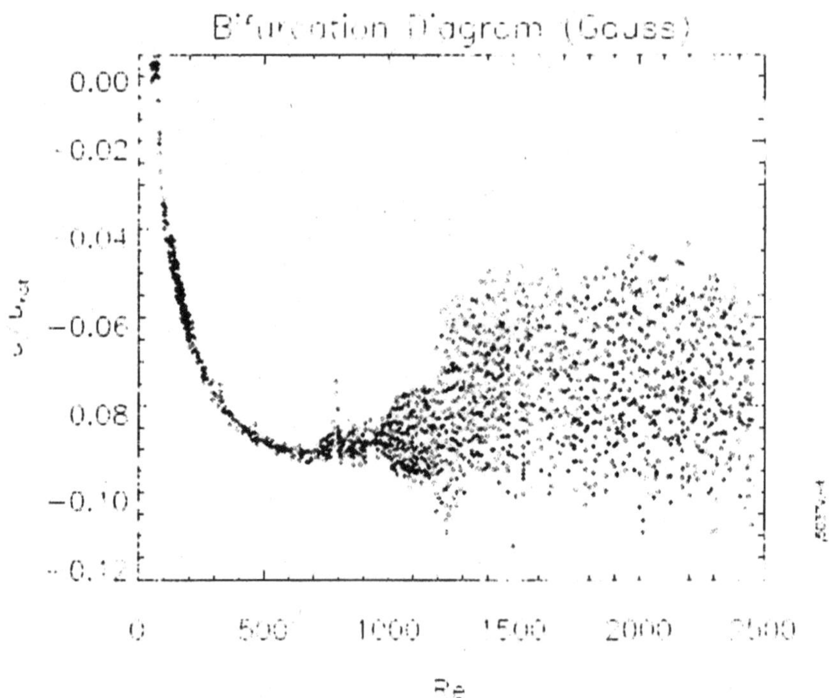

Figure 14: Bifurcation diagram in the Taylor Couette system as a function of the Reynolds number. The meridional velocities are normalized by the velocity of rotation (η=0.5, Γ=3.97, z=0.37H, $(r-R_1)/(R_2-R_1)$=0.5)

But to detect the nonlinear behaviour during the route into chaos, in addition it is necessary to use nonlinear signal processing techniques to analyse the time series. Therefore, an alternative way to investigate the bifurcations and the nonlinear dynamics is to reconstruct the underlying attractors which represent the dynamic behaviour. This is done by the method of time delayed coordinates introduced by Takens (1980). The appropriate time delay τ and the embedding dimension dimE are estimated by the calculation of the fill factor and the integral local deformation (Buzug & Pfister, 1992) and singular value decomposition (Broomhead & King, 1986). The reconstructed attractor is projected onto an orthonormal basis that is calculated from the covariance matrix of the time-delayed time series. This can be interpreted as an optimal stretching of the attractor on the primary axes of an dimE -dimensional hyper ellipsoid containing the averaged spatial extension of the attractor in the phase space. It is then possible to apply an estimation calculation of the Lyapunov exponents onto the attractor. The sign and magnitude of the greatest exponent is providing information about the dynamic behaviour of the system. The Lyapunov exponent spectrum is calculated by a linear approximation of the local flow in the phase space (Holzfuss, 1987). A detailed description of the non-linear behaviour during the route into chaos in the wide gap spherical Couette flow with reconstructed attractors and calculated Lyapunov exponents can be found in the work of Wulf et al. (1997).

6.3 Time Series Analysis of Rossby Waves

To detect the dynamic behaviour of the occuring Rossby waves in the rotating annulus using linear signal processing techniques the most common and very usefull way is to analyse the time series of a representative velocity component, which is in the rotating annulus experiment the radial one, and to construct the power spectra and the autocorrelation function. The spectrum gives a measure of the amount of power in a given frequency band over a selected frequency range. The autocorrelation function for a peridic signal for example is itself periodic and can often give a less confusing representation of the data than the power spectrum. It therefore provides a useful complementary representation of the data. Irregularity in the time series give rise to a decay in the autocorrelation function and the rate of decay gives a measure of the degree of irregularity. However, this method is linear and therefore fails when chaotic motion occurs. The extraction of quantitative estimates from the autocorrelation function can be problematical and it is usually used as a complement to the power spectrum.

The results of time series analysis due to the LDV measurements in the rotating annulus heated from within are summarized in this chapter for slightly supercritical parameters, where Rossby waves exist. The diagrams depicted in figure 15 contain:

- o velocity time series
- o pictures of flow visualization
- o power spectra
- o autocorrelation function

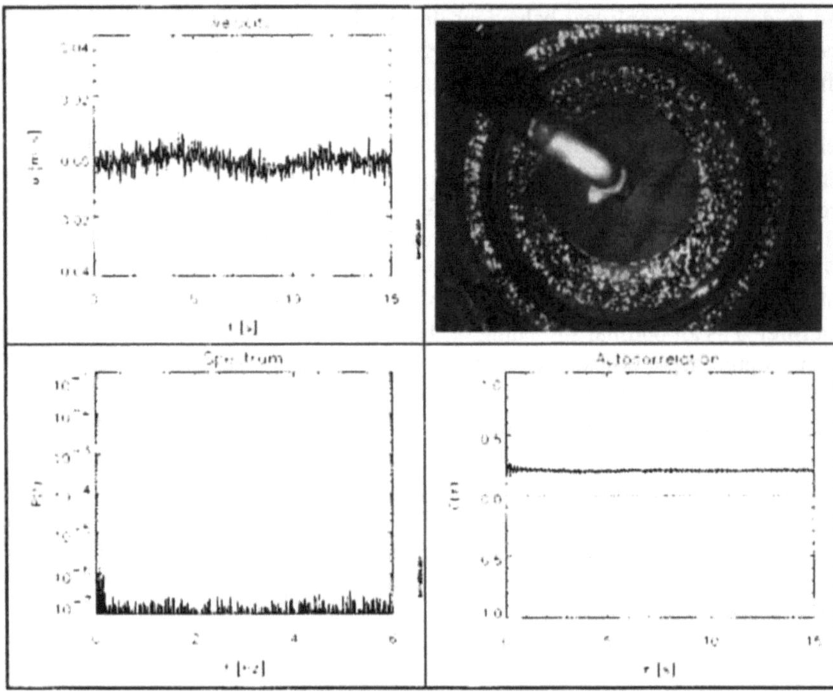

Figure 15 (a): Basic Flow: m = 0, Ro = 17.2, Ta = $2.45 \cdot 10^5$

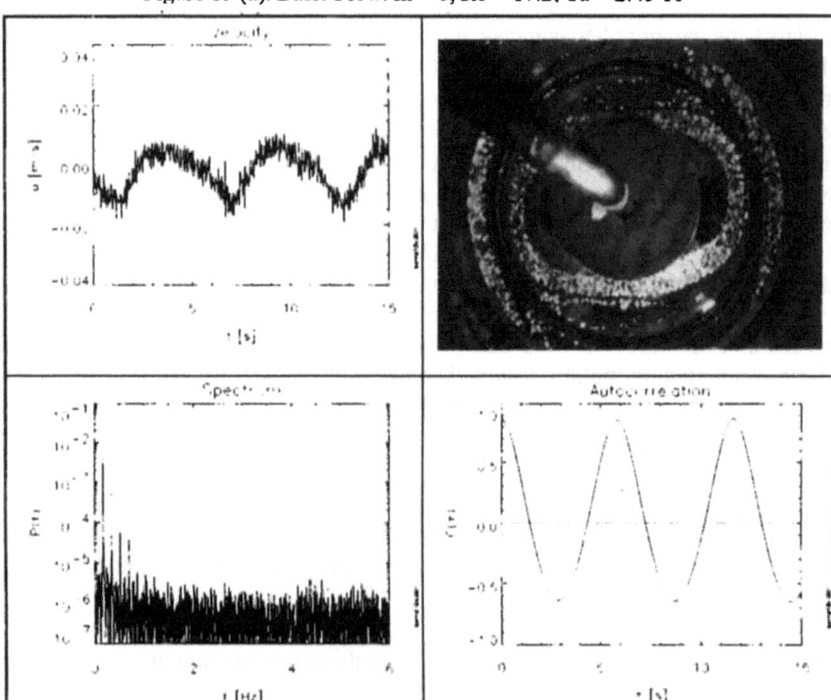

Figure 15 (b): Rossby Waves: m = 2, Ro = 2.74, Ta = $1.56 \cdot 10^6$

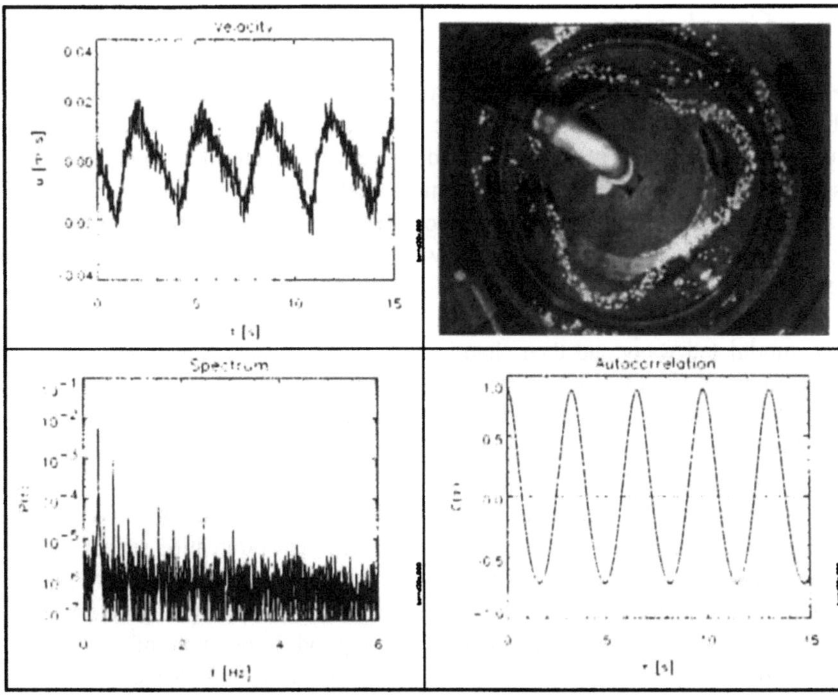

Figure 15 (c): Rossby Waves: m = 3, Ro = 1.40, Ta = $3.01 \cdot 10^6$

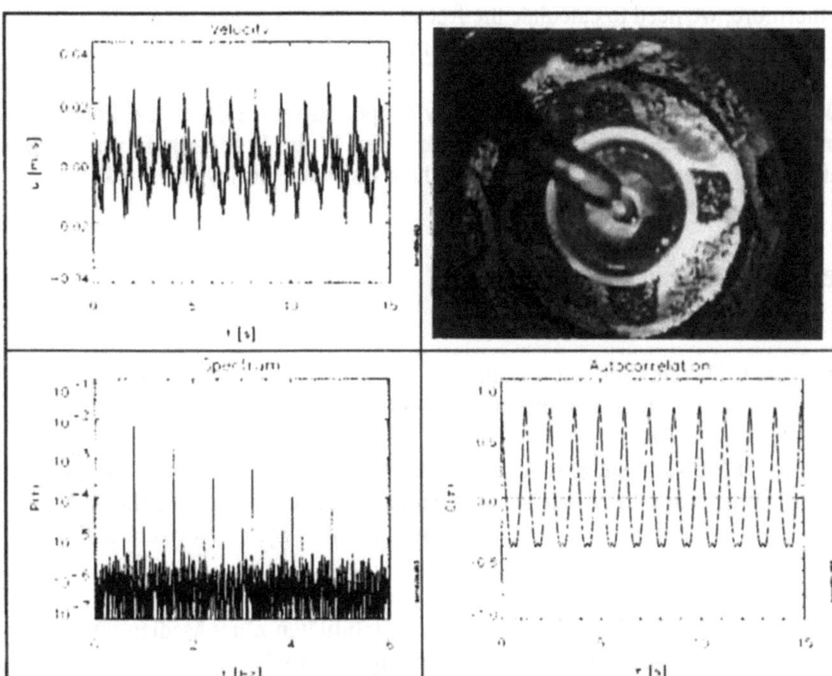

Figure 15 (d): Rossby Waves: m = 4, Ro = 0.48, Ta = $8.97 \cdot 10^6$

7 Conclusion

We have considered some characteristics of bifurcation scenarios during the laminar turbulent transition in rotating fluid flows. In contrast to previous investigations, we could detect a transition region in the wide spherical Couette flow, where instabilities in the form of spiral waves occur, which break the spatial symmetry. The basic flow in the wide gap spherical Couette flow bifurcates via different Hopf bifurcations into chaos. In contrast, the basic flow between two concentric cylinders bifurcates via a pitchfork bifurcation into the Taylor vortex flow and then depending on the radius ratio the second bifurcations are different. In small gaps the flow becomes wavy, modulated wavy and then chaotic with increasing the Reynolds number. In wide cylindrical gaps the Taylor vortex flow becomes disturbed and then chaotic. The LDV-technique could be applied successfully for measurements of the azimuthal and meridional velocity components. In this way, the time-dependent behaviour of the different instabilities in rotating systems during the transition to turbulence could be obtained. The method of counting minima and maxima of the oscillating velocities was successfully applied to obtain the different bifurcation scenarios. The systems show the typical rich variety of routes to chaos, e.g. the periodicity, quasi-peridicity and stochastic regions. Simultaneous measurements of the velocity components and flow visualization studies give a good impression of the spatial-temporal development of the flows during the transition to turbulence. For a complete description of these transitions, however, we need to conduct simultaneous high-resolution LDV-measurements of the fully three-dimensional velocity vector. Furthermore, we need to calculate the fractal dimensions to analyze the properties of the dissipative non-linear dynamical systems.

Acknowledgements. This work is partly supported by the Deutsche Forschungsgemeinschaft (DFG). The authors thank Peter Wulf, Wolfgang Beyer, Rüdiger Maier, Antje Brucks and Bernd Sitte for their assistence during this work.

References

Andereck, C.D., Liu, S.S. and Swinney, H.L. (1986): Flow regimes in a circular Couette system with independently rotating cylinders. J. Fluid Mech., vol. 164, 155-183

Belyaev, Yu.N., Monakhov, A.A., Scherbakov, S.A. and Yavorskaya, I.M. (1984): Some routes to turbulence in spherical Couette Flow. in: Kozlov,V.V. (ed.): Laminar-Turbulent Transition. IUTAM-Symp. Novosibirsk/USSR, Springer

Broomhead, D.S. & King, G.P. (1986): Physica D 20, 217

Bühler, K. (1985): Strömungsmechanische Instabilitäten zäher Medien im Kugelspalt. VDI-Berichte, Reihe7: Strömungstechnik Nr.96

Bühler, K. and Zierep, J. (1984): New secondary instabilities for high Re-number flow between two rotating spheres. in: Kozlov,V.V. (ed.): Laminar-

Turbulent Transition. IUTAM-Symp. Novosibirsk/USSR, Springer, Berlin, Heidelberg, New York, Tokyo

Bühler, K. and Zierep, J. (1986): Dynamical instabilities and transition to turbu lence in spherical gap flows. in: Comte-Bellot, G. and Mathieu, J.: Advan ces in turbulence. Proc. 1st Europ. Turb. Conf., Lyon, France, 1-4 July, Springer, Berlin,Heidelberg, NewYork,London, Paris, Tokyo

Buzug, Th. & Pfister, G. (1992): Physical Review A 45, 10

Buzug, Th., v. Stamm, J. and Pfister, G. (1992): Fractal dimensions of strange attractors obtained from the Taylor-Couette experiment. Phys.A, 191, 559

Buzug, Th., v. Stamm, J. and Pfister, G. (1993): Characterization of period-doubling scenarios in Taylor- Couette flow. Physical Review E, 47, no. 2, 1054-1065

Eckmann, J.P. and Ruelle, D. (1992): Fundamental limitations for estimating di mensions and Lyapunov exponents in dynamical systems. Physica D 56

Egbers, C.(1994): Zur Stabilität der Strömung im konzentrischen Kugelspalt. Dissertation, Universität Bremen

Egbers, C. and Rath, H.J. (1995): The existence of Taylor vortices and wide-gap instabilities in spherical Couette flow. Acta Mech. 111, 3-4, 125-140

Egbers, C. & Rath, H.J. (1996): Developments in Laser Techniques and Applicati ons to Fluid Mechanics (Eds.: R.J. Adrian, D.F.G. Durao, F. Durst, M.V. Heitor, M. Maeda, J.H. Whitelaw), Springer,45-66

Fechtmann, C., Wulf, P., Egbers, C. & Rath, H.J.; 1996; LDA-Messungen zum chaotischen Verhalten der Strömung im konzentrischen Kugelspalt; Lasermethoden in der Strömungsmeßtechnik, 5. Fachtagung der GALA, 11.-13. Sept. Berlin, vol. V, pp 17.1-17.7

Fenstermacher, P.R., Swinney, H.L. and Gollub, J.P. (1979): Dynamic instabilities and the transition to chaotic Taylor vortex flow. J. Fluid Mech., 94, part1, 103-128

Gollub, J.P. and Swinney, H.L. (1975): Onset of turbulence in a rotating fluid. Phys. Rev. Lett. 35

Hide, R. and Mason, P.J. (1975): Sloping convection in a rotating fluid. Adv. Phys. 24, 47-100

Holzfuss, J. (1987): Doctoral thesis, Universität Göttingen,

Landau, L.D. (1944): On the problem of turbulence. C. R. Acad. Sci. USSR 44, 311

Liu, M., Blohm, C., Egbers, C., Wulf, P. and Rath, H. J. (1996): Taylor vortices in wide spherical shells.; Phys. Rev. Lett., vol. 77, No. 2, pp 286-289

Marcus, P.S. (1988): Numerical simulation of Jupitor's Great Red Spot. Nature 338, 693-696

Mullin, T. (1993): The nature of chaos, Oxford Science Publications

Munson, B.R. and Menguturk, M. (1975):Viscous incompressible flow between concentric rotating spheres. Part 3: Linear stability and experiments. J. Fluid Mech., vol. 69, 705-719

Nakabayashi, K. and Tsuchida,Y. (1988): Spectral study of the laminar-turbulent transition in spherical Couette flow. J. Fluid Mech., vol. 194, 101-132

Newhouse, S., Rouelle, D. Takens, F. (1978) Occurence of strange axiom attractors near quasiperiodic flow on T^m, $m \geq 3$, Commun. Math, Phys. 64

Press, W.H., Teukolsky, S.A., Vetterling, W.T. and Flannery, B.P. (1994): Numerical recipes in C, Cambridge Univ. Press

Rouelle, D. and Takens, F. (1971): On the nature of turbulence. Commun. Math. Phys. 20, 167-192

Sawatzki, O. und Zierep, J. (1970): Das Stromfeld im Spaltzwischen zwei konzentrischen Kugelfl„chen, von denen die innere rotiert. Acta Mech. 9, 13-35

v. Stamm, J.,Buzug,Th. and Pfister,G. (1993): Frequency locking in axisymmetric Taylor-Couette flow. submitted to Physical Review E.

Takens, F. (1980): Lecture Notes in Mathematics, 898, Springer

Taylor, G.J. (1923): Stability of a viscous liquid contained between two rotating cylinders. Phil.Trans. A223, 289-293

Wimmer, M. (1981): Experiments on the stability of viscous flow between two concentric rotating spheres. J. Fluid Mech., vol.103, 117-131

Wulf, P., Egbers, C. and Rath, H.J. (1997): Routes to chaos in wide gap spherical Couette flow. Phys. of Fluids, submitted

Yavorskaya, I.M., Belyaev, Yu.N. and Monakhov, A.A. (1975): Experimental study of a spherical Couette flow. Sov. Phys. Dokl., vol. 20, 4, 256-258

Yavorskaya, I.M., Belyaev,Yu.N., Monakhov, A.A., Astaf'eva, N.M, Scherbakov, S.A. and Vvedenskaya, N.D. (1980): Stability, non-uniqueness and transition to turbulence in the flow between two rotating spheres. IUTAM-Symposium, Toronto

LIF-Diagnostics for the Investigation on the Chemical Structure of Diffusion Flames Burning Under Microgravity

J. König, Chr. Eigenbrod, H.J. Rath

Center of Applied Space Technology and Microgravity, ZARM, University of Bremen, Am Fallturm, D-28359, Germany

Abstract

Laser induced fluorescence (LIF)- diagnostics has been developed for the use in microgravity combustion experiments in the Bremen Drop Tower. By means of an KrF-Excimer Laser as UV-excitation source, different LIF- and LIPF-diagnostic methods are beeing applied to investigate on the chemical structure of igniting and burning single fuel drenched porous spheres nonintrusively. The two-dimensional OH-radical concentration field, identifying the place and strength of reaction, was detected with a temporal resolution of 4ms. By means of an alternating fast wavelength switching, the temperature distribution in the vicinity of a reaction zone adjacent to a methanol drenched porous sphere was measured utilizing two line OH-LIPF.
Converting the laser wavelength of 248nm by means of Raman-shifters, the visualisation of important cool flame intermediates by LIF is under preparation for the use in microgravity experiments.

Keywords: Laser-induced fluorescence, Chemically reacting flow, Microgravity, Droplet combustion

1. Introduction

The Experimental investigation of flow fields prominently requires the availability of non intrusive, physically nondisturbing measurement systems. Especially researching on the behavior of very weak flow phenomena, or on flow conditions close to the onset of instable or bifurcated behavior, any type of interference of the utilzed measurement tool with the object beeing observed can affect the obtained data in an inacceptable way. The issue becomes even harder to be solved, if the observed flow is chemically reacting and care has to be taken neither to disturb the flow pattern physically nor disturbing the character of the chemical reaction by means of sensors or probe sampling. Considering most of the flows of interest beeing nonsteady and geometrically at least two-dimensional, the choice of adequate and precise diagnostics is one of the great issues researching on those flows experimentally.

Microgravity is a boundary condition that is especially favorable for all experiments, which are focussing on the investigation of effects that are typically weaker than those caused by sedimentation, or convection driven by density gradients. Due to that, the absence of gravity prevents in particular cases the flow from establishing a complex three dimensional behavior, easing a numerical modelling.
Investigating on the combustion of single fuel droplets, microgravity is a powerful tool to prevent the unsteady thermal convection driven flow from veiling the mainly diffusional controlled fundamental processes of chemical reaction [1,2]. Additionally, analysing the flow conditions in real type fuel sprays, it was found that the typically very small droplets follow the flow pattern easily and behave almost like in a quiescent ambience. Since the burning lifetime is very short a bouyant convective flow can not be established combustion takes place almost like under microgravity condition [3,4]. Thus, investigating on droplet combustion in the absence of gravity, the obtained results do either deliver a fundamental understanding of diffusion combustion of isolated liquid fuel spheres and important insights to the combustion of fuel sprays as well. As for the affore mentioned flows, microgravity droplet combustion represents a typical domain of a highly sensible flow, coinciding with chemical reaction in the gas phase combined with the requirement of temporally and spatially resolved measurements.
Searching for adequate diagnostic means beyond classical optical methods like schlierenphotography, shadowgraphy, interferometry or regular high speed cinematographic observation, at the drop tower Bremen a UV-Excimer Laser based diagnostic system has been developed to apply laserspectroscopic tools to experiments in microgravity environment [5, 6].

2. Setup and specifications of the laserdiagnostics (LDS) at the Bremen drop tower

A UV-Excimer Laser in Oscillator-Amplifier configuration delivers a tunable, narrowband high energy light source at 248nm (KrF) wavelength, pulsed with a repetition rate of 250 Hz. Due to its large size, high weight and power consumption, the laser is attached at the top of the vaccum tube in the tower. After beeing expanded to a size of 25x 25mm by means of a telescope, the UV-beam exhibits an almost constant shape following the falling capsule for 110 meters distance. Before entering the vacuum tube and beeing mirrored towards the top end of the falling capsule, the UV beam is overlayed coaxially with a HeNe pilot laser. The pilot laser is used to correct allignment- and misplacement errors of the UV-beam relative to the capsule during free flight in a so called "Pointing Assembly". That Pointing Assembly, mounted at the top end inside the capsule, mirrors the beam to a well defined spot in the experiment chamber during the complete flight. The entire setup of the components is depicted in

figure 1. Even though a change of the operational wavelength of 248nm to 193nm, 308nm or 352nm is possible, the large number of specially coated optical components beeing applied between laser and experiment promotes the conversion of the wavelength, if nessecary, inside the capsule. Due to the high pulse energy of the beam, entering the capsule with at least 250mJ per pulse at 250Hz repetition rate, conversion techniques with even low degrees of efficency are suitable.
The drop bus is equipped with a fully autonomous data aqquisition system, that, lacking a wire connection between laser and drop capsule, generates an appropriate trigger regime for the gated two stage image intensifier and the high speed video camera by the respective preceeding laser. The high-speed camera is a modified standard Dalsa camera, operating at typically 250Hz with

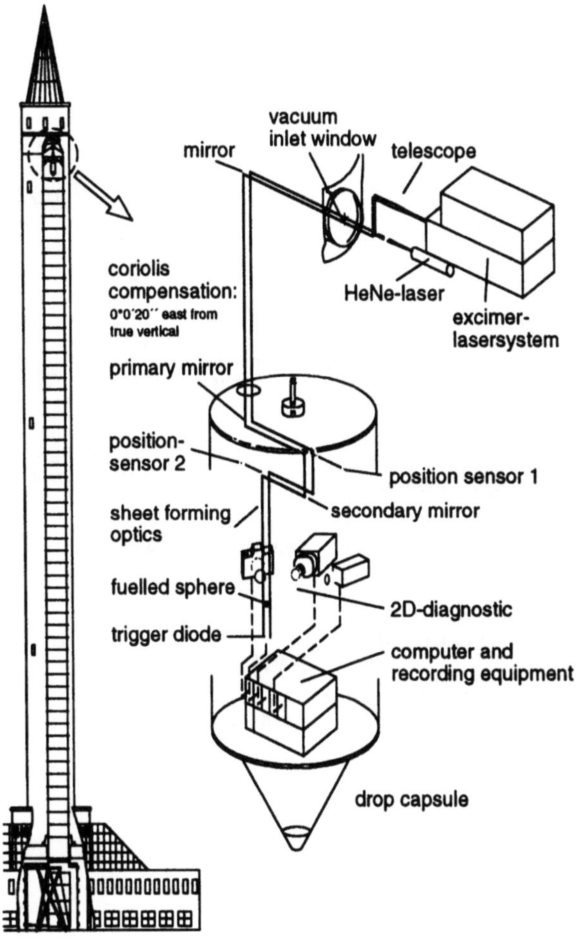

Fig. 1: Outline of the UV-laserdiagnostic at Bremen Drop Tower

256x 256 pixels field of frame and a dynamic range of 8 bit. The video data are stored in real time on a 96Mbyte D-Ram, offering the recording of 1512 full frame pictures or 6 seconds of high speed video data. By selecting appropriate filters in the optical path of the intensified camera and readjusting the timing, diverse laser diagnostic methods can be used inside the drop capsule. Figure 2 shows the drop bus equipped with the complete infrastructure used for the application of UV-laserdiagnostics.

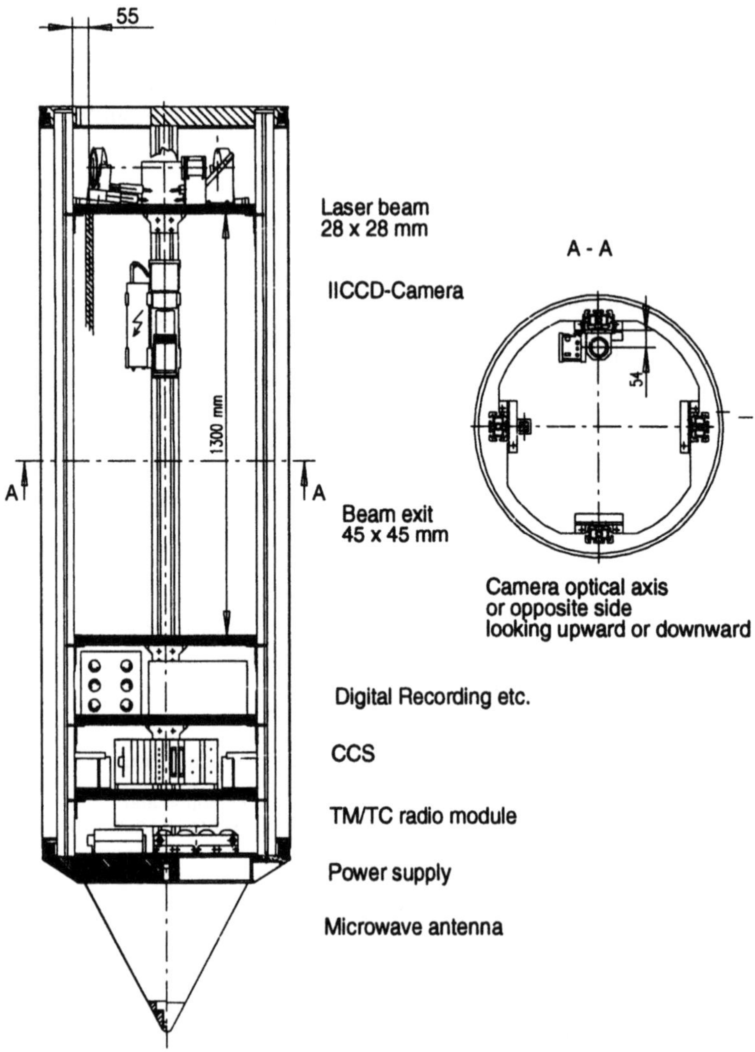

Fig. 2: Drop capsule including infrastructure for experiment control and laserdiagnostic

3. Application of OH-LIPF to Microgravity Droplet Combustion
3.1. Single Wavelength excitation

Observing the ignition and combustion of fuel droplets under microgravity, the methods of data acqusition are usually limited to detect the image of the irradiation and absorption of the flame, or to measure the integral optical properties in the gas phase by means of e.g. schlieren- or interferometric techniques. Information on the chemical structure gain a strong interest, as they are of special importance as input data for numerical simulation of droplet combustion and the basic understanding of diffusion combustion.
At Bremen Drop Tower the development of the concentration field of OH-radicals in the gas phase around igniting and burning droplets could be detected two-dimensionally and highly temporally resolved for various fuels. The excitation wavelength was tuned to the $P_2(8)$ line in the electronic $A^2\Sigma^+ \rightarrow X^2\Pi$ transition [5,6] which yields a predissociation flourscence signal that is almost independant of the temperature (above 1000K) and unaffected by collisional quenching. To facilitate the application of this diagnostics, isolated fuel drenched porous spheres of 5.6mm in diameter, were used instead of small all-fuel droplets. The experimental configuration is shown in figure 3. The fuel drenched porous sphere is shielded by an aperture to prevent direct illumination and light scattering caused by the UV-laser lightsheet.

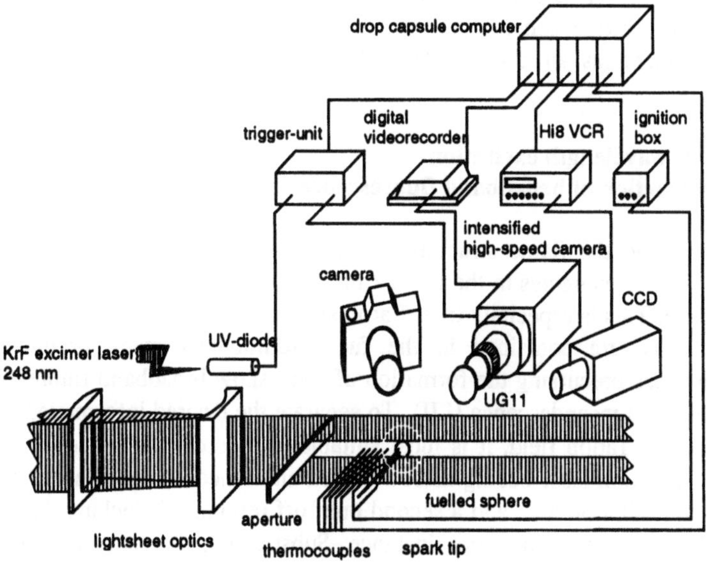

Fig. 3: Experimental setup for LIF-diagnostic on a burning fuel drenched porous sphere

Figure 4 shows a sequence of three images of the OH-radical field beeing formed around a porous sphere using methanol as fuel. In the presented case, the sphere was ignited 200ms after releasing the drop capsule. By analyzing the fluorescence images, which show the typical sperical shape of a microgravity droplet flame, the position of the reaction zone can easily be determined [7,8]. Sideeffects, like the decompostion of product water to OH-radicals in a two photon absorption process is supposed to be of minor importance. Using the advantage of the spherical symmetry, one dimensional simulations for droplet combustion can be applied to calculate the transient ignition and burning behavior using advanced multistep reaction kinetics.

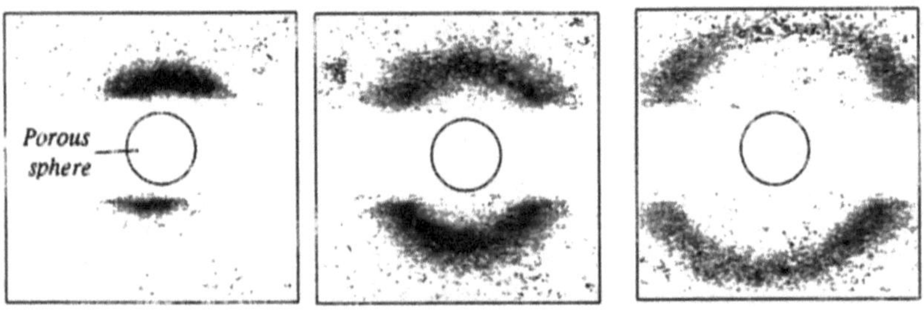

t_0+ 25ms t_0+ 200ms t_0+ 1000ms

Fig.4 Sequence of OH-fluorescence images 25ms, 200ms and 1000ms after ignition, sphere diameter 5.6mm, fuel: methanol; ambient gas: air, p= 1bar, T= 293K

3.2. Multi wavelength excitation
3.2.1. Subtraction of broadband fluorescence

Investigating on the flame structure of hydrocarbon fuels by means of LIF, especially at wavelenghtes in the far UV the fuel shows considerable broadband fluorescence. This is especially true if aromatic components, even in small trace concentrations, are contained in the fuel. Additionally many products, in particular those precursing the formation of soot, show broadband fluorescence or laser induced incandescence (LII). To separate the desired information about the OH-concentration field, it is most suitable to record a fluorescence image with the excitation laser beeing narrowbanded and tuned "on resonance" to the OH-radical excitation line, and a second one "off resonance", lacking the signal of OH-fluorescence at the same instance. Substracting the recorded pictures, purely the image of the OH-concentration field remains [7]. Figure 5 shows an example applying this technique to a burning n-heptane drenched porous sphere under microgravity condition at t = 400ms after ignition. Since there is only a single laser source, a fast wavelengh switching unit was implemented to the narrowband unit of the excimer laser [11]. The littrow grating, which represents the optical component tuning the laser wavelength in the oscillator discharge tube, is mounted on an oscillating piezzo actuator, which allows to change the wavelengh in different modes periodically. Applying the alternating on-, off-resonance fluorescence excitation diagnostic, a puls repetition rate of 250Hz yields a 125Hz repetition rate of two frame sequences.

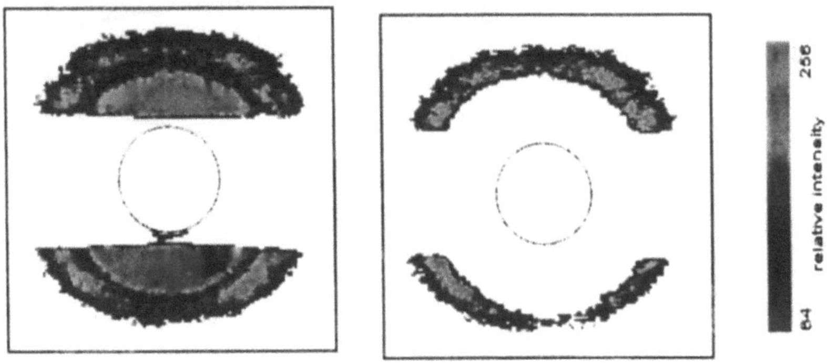

A) B)

Fig. 5: LIPF- images of an n-heptane drenched sphere burning under microgravity condition. A: OH- plus broadband fluorescence, B: OH-fluorescence obtained by on-, off- resonance image subtraction.

Using the two wavelength exitation technique, the effective exposure time for one resulting image is 4ms, compared with typically tens of nanoseconds (which refers to the lifetime of induced fluorescence) when using a single wavelength excitation technique.

Considering the transient development of the flame, except close to ignition, beeing slow compared to a time scale of 4ms, the resulting data can be used to obtain a temporally and spacially resolved OH-radical structure in flames, even if the OH-LIF is obscured by laser induced side effects. Those data are highly suitable to validate numerical calculations of droplet combustion including chemical reactions.

3.2.2. Transient temperature field measurement by means of two wavelength OH-LIPF

Using the electronic $A^2\Sigma^+ \rightarrow X^2\Pi$ transition of the OH-radical, there are six lines that can distinctly be excited within the tunable bandwidths of an excimer laser operated at 248nm, yielding a strong LIPF signal [8,12]. Each individual line shows a typical temperature dependant signal strengh, which is mostly due to the population density distribution of the respective electronic ground state level. Figure 6 shows the temperatur dependency of the most prominent fluorescence lines of that electronic transition. Using the alternating two wavelength excitation switching technique described in 3.2.1., two dimensional temperature fields can be determined in areas of considerable OH-concentration, i.e. in the vicinity of the reaction zone.

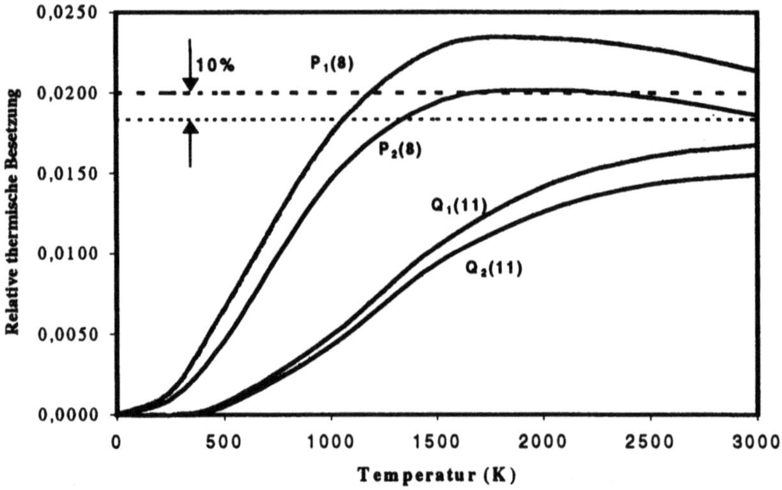

Fig. 6: Population of the $P_1(8)$, $P_2(8)$, $Q_1(11)$ and $Q_2(11)$ line in the electronic $A^2\Sigma^+ \rightarrow X^2\Pi$ transition as a function of temperatur [ ]

Figure 7 illustrates the generation of a temperature distribution image obtained by two preceeding LIPF images exciting the $P_2(8)$ and $Q_1(11)$ line. Above 1300K the $P_2(8)$ line can be considered as temperature independant, whereas the LIPF signal of the $Q_1(11)$ contains a temperature information. The images show a methanol drenched porous sphere burning under microgravity, 400ms after ignition. The underlying mathematical procedure is described in [12,13]. Actually, to employ this technique in quasi single shots to obtain 125Hz temporally resolved information, the source images have to be of high quality and high signal to noise ratio. To improve the quality of the raw data, a third laser shot with the laser tuned "off-line" is usually required to subtract the background signal. As can be seen from Figure 7 the calculation of the temperature field becomes inappropriate if the gas temperature is close or below 1300K and if the OH-concentration is to low to give a considerable signal strength.

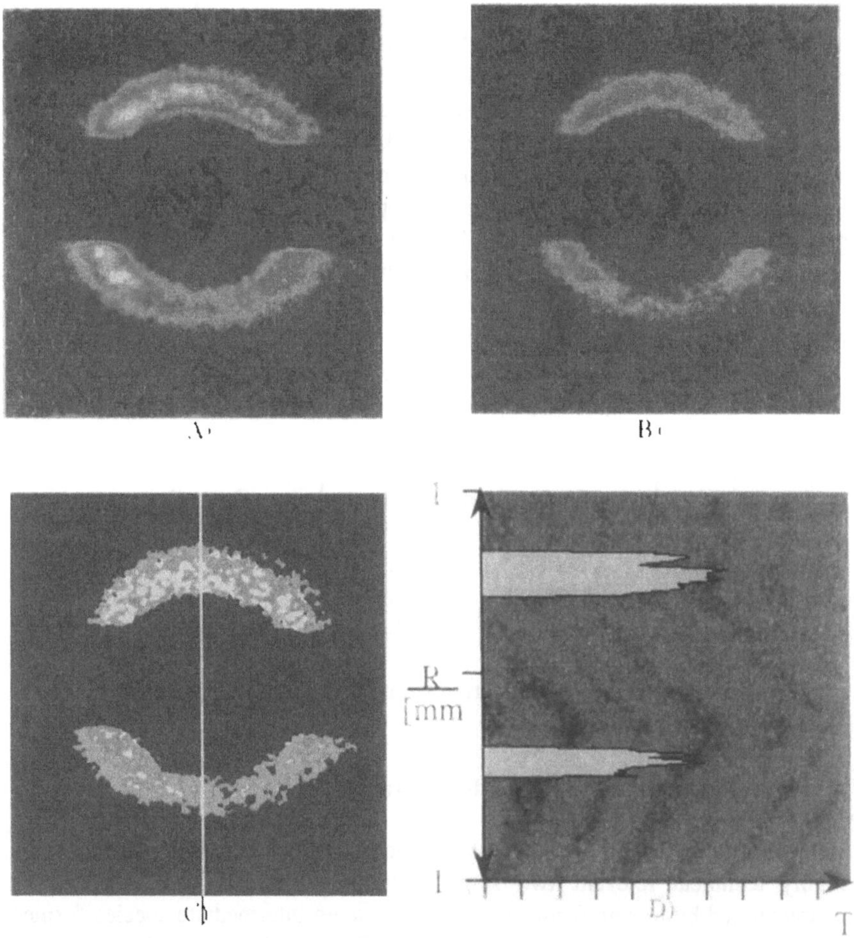

Fig.7: Generation of a temperature image out of two succesively taken (dt= 4ms) OH-LIPF images. The observed object is a methanol drenched porous sphere 400ms after ignition, burning under microgravity condition.
A) OH-LIPF, line $P_2(8)$ excited, B) OH-LIPF, line $Q_1(11)$ excited,
C) calculated temperature image, D) temperature profile at cutline

4. Detection of combustion intermediate species in microgravity combustion experiments using Raman-shiftet excitation wavelengths

Using an excimer laser as high energy source for laser based diagnostics, there are several opportunities to convert the excitation wavelength. The most flexible solution is to do this conversion in the proximity of the experimental zone, because only few optical components between converter and observed object have to be changed regarding there wavelength depending coatings or properties. For high energy applications a favorable way of wavelength conversion is, except pumping a dye laser stage, inserting a Raman-active media into the beam path of the excimer laser [14,15]. Induced by stimulated Raman scattering, the emitted light shows strong lines of several stokes and antistokes Raman shifted orders. Depending on the application, these Raman lines can be generated with a similar bandwidth like the initial stimulating laser light. After leaving the Raman converter, the emitted light can be spectrally seperated by means of diffractive optics. The efficience of the conversion process is mainly a function of the density of the Raman media, the geometry of the converter, and the stokes-order of the selected wavelength. In Table 1 some examples of Raman shiftet wavelength for potential future use in the drop capsule are listet.

Media	Line	Wavelength [nm]	Efficiency	Potential use
H_2	AS 1	225nm	3,5%	NO $(A^2\Sigma^+ \rightarrow X^2\Pi)(0\text{-}0)$
H_2	2 S	313nm	16,0%	Form-, Acetaldehyde
D_2	4 S	353nm	n.n.	Formaldehyde

Table 1: Output wavelengths converted from 248nm in a Raman cell

4.1. LIF of formaldehyde as cool flame tracer

Investigating on the ignition and combustion of hydrocarbon fueled droplets, the highly technical relevant two stage ignition process of n-alkanes can be characterized by tracing formaldehyde as the main intermediate species formed during cool flame burning [16,17,18]. Performing these experiments under conditions of microgravity, the results can easily be used to gain input data for accompanying numerical simulations including chemical kinetics for either low- and high temperature reactions.

Figure 8 shows a sequence of a proceeding two stage ignition of a n-heptane drenched porous sphere (diameter 5,6mm) rapidly inserted into a furnace with 590K in 0.3 MPa ambient air. The images show qualitative LIF of formaldehyde excited at 353,2nm, indicating the place and time of the two stage ignition. Actually, these experiments have been performed under normal gravity, qualifying the furnace for the use in the drop capsule. The laser light sheet is adjusted to the space directly underneath the n-heptane drenched porous sphere, since, under the given condition, the ignition process takes place at the lower side of the sphere. The position of the sphere is visible in the top center of each frame. Figure 8a shows the concentration field of formaldehyde at the instant of cool flame ignition 420ms after insertion. Figure 8b shows the formaldehyde distribution of a fully developed cool flame, burning directly below the porous sphere 850ms after insertion. At t = 900ms hot ignition occurs, rapidly consuming the formaldehyde molecules.

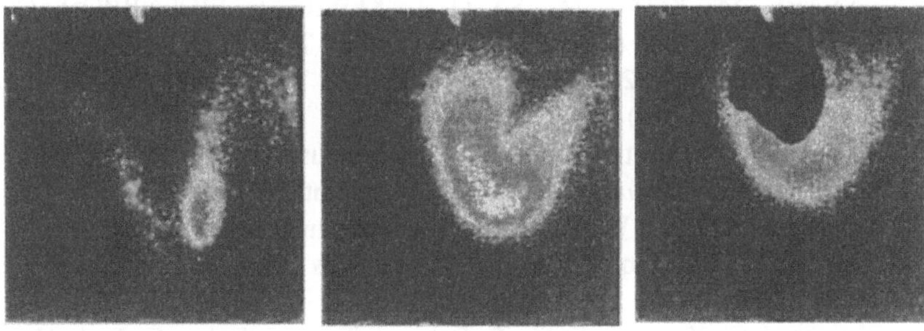

t_0 + 420ms t_0 + 850ms t_0 + 900ms

Fig. 8: Sequence of formaldehyde-LIF of a selfigniting n-heptane drenched porous sphere, normal gravity, p= 3bar, T=590K, ambient gas: air.

5. Conclusion

Different two-dimensional laserspectroscopic methods have been applied to microgravity combustion experiments in the drop tower Bremen. Due to special modifications of the laser system, various techniques that require succesive laser pulses of different narrowbanded wavelengths can be utilized. Laser induced fluorescence techniques, measuring the OH-radical distribution, the temperature field in the vicinity of the flame, as well as the concentration of formaldehyde concentration as an important tracer of low temperature reactions were tested succesfully. The obtained results are very promising to explore the chemical structure of various combustion phenomena under microgravity.

Aknowledgement

The authors like to thank the German Space Agency DARA (now DLR) for their kind support of the presented work. (DARA 50 WP 9251 und 50 WM 9448).

References

1. C.K. Law, G.M. Faeth: "Opportunities and challanges of combustion in microgravity", *prog. energy and comb. sci.*, 20, 65, London (1994)
2. TANABE, M., KONO, M., SATO, J., KÖNIG, J., EIGENBROD, CH., RATH, H.J.,: Effects of Natural Convection on Two Stage Ignition of an n-Dodecane Droplet, *25th Symp.(Intl.) Comb.*, The Combustion Institute, Pittsburgh,455 (1994)
3. Bachalo, W.D.:"Injection, dispersion, and combustion of liquid fuels, *25th Symp.(Intl.)Combustion*, The Combustion Institute,333 (1994)
4. Annamalai, K.;Ryan, W.:"Interactive processes in gasification and combustion. PartI: Liquid drop arrays and clouds", *Prog. Energy Comb. Sci*, 18, 221, (1992)
5. Eigenbrod, C., König, J., Bolik, T., Behrens, T., Dinkelacker, F., Rath, H.J., Albrecht, H., Müller, D., Schröder, T., Triebel, W.:"Development of an UV-laser Diagnostic System for Combustion Research under Zero-gravity at Drop Tower Bremen, *7th Int. Symp. Appl. of Laser Tech. to Fluid Mech.*, Lisboa (1994)
6. Eigenbrod, C., König, J., Bolik, T., Renken, H., and Rath, H.J.: First results from non-intrusive laser diagnostic system for combustion research at Bremen drop tower, *Microgravity sci. technol.* VIII/2: 134-136 (1995).
7. Kychakow, G., Howe, R.D., Hanson, R.K.: "Use of planar laser-induced fluorescence for the study of combustion flowfields, *AIAA No.83-1361*, Seattle (1983)
8. Kohse-Höinghaus, K.:"Laser techniques for the quantitative detection of reactive intermediates in combustion systems", *Prog. Energy Comb. Sci.*,20, 203, London (1994)
9. König, J., Eigenbrod, C., Tanabe, M., Renken, H, Rath, H.J.: ''Characterization of spherical hydrocarbon fuel flames: laser diagnosis of the chemical structure through the OH-radical, *26th int. symposium on combustion*, Neapel (1996).
10. Renken, H., Bolik, T., Eigenbrod, C., König, J., Rath H.J.:''Two-dimensional UV-laserdiagnostic by high-speed imaging for microgravity combustion research at Bremen drop tower *AIAA 97-0241*, 35th Aerospace Sciences meeting and exhibit, Reno, (1997).
11. Grebner, D., Müller, D., Triebel, W.:"Fast wavelength switching of narrow-band excimer lasers", *Rev. Sci. Instrum.*, 68 (8), 2965, (1997)
12. Arnold, A.:"Entwicklung von abstimmbaren Mehrfrequenz-Excimer-Lasern zur Bestimmung von Konzentrations- und Temperaturfeldern in turbulenten technischen Flammen", *Disserationsschrift*, Heidelberg, 1992
13. Hinrichs, O.:"Zweidimensionale Temperaturfeldbestimmung an Methanol-diffusionsflammen in Mikrogravitationsexperimenten mittels 2-Linien OH-LIPF", *Masterthesis*, Bremen, (1997)

14. Dreier, T., Dreizler, A., Wolfrum, J.: "The application of a raman-shifted tunable KrF excimer laser for laser-induced fluorescence combustion diagnostics", *Applied Physics B*, 55, 381, (1992)
15. Schoulepnikoff, V. Mitev, V. Simeonov, B. Calpini, H. a.d. Bergh:" Experimental investigation of high-power single-pass Raman shifters in the ultraviolet with Nd:YAG and KrF lasers, *Applied Optics*, 36, 21, 5026, (1997)
16. Bäuerle, B., Warnatz, J., Behrendt, F.:"Time-resolved investigation of hot spots in the end gas of an S.I. Engine by means of 2-D Double pulse LIF of formaldehyde, *26th Symp. (Intl.) on Comb.*, The Combustion Institute, Pittsburgh, 2619 (1996)
17. Müller, U.C., Peters, N.; Linan, A.:"Global kinetics for n-heptane ignition at high pressures", *24th Symp. (Intl.) Comb.*, The Comb. Institute, Pittsburgh, 777 (1992)
18. Tanabe, M. Kono, M., Sato, J., König, J., Eigenbrod, C., Dinkelacker, F., Rath, H.J.,:"Two stage ignition of n-heptane isolated fuel droplets, *Combust. Sci tech.* 108:103 (1995).

MIX
Papier aus verantwortungsvollen Quellen
Paper from responsible sources
FSC® C105338

If you have any concerns about our products,
you can contact us on
ProductSafety@springernature.com

In case Publisher is established outside the EU,
the EU authorized representative is:
Springer Nature Customer Service Center GmbH
Europaplatz 3, 69115 Heidelberg, Germany

Printed by Libri Plureos GmbH
in Hamburg, Germany